Forests, Business and Sustainability

Forests are under tremendous pressure from human uses of all kinds, and one of the most significant threats to their sustainability comes from commercial interests. This book presents a comprehensive examination of the interactions between the forest products sector and the sustainability of forests.

It captures the most current sustainability concerns within the forestry sector and various sustainability-oriented initiatives to address these. Experts from around the world analyze interconnected topics including market mechanisms, regulatory mechanisms, voluntary actions, and governance, and outline their effectiveness, potential, and limitations. By presenting a novel overview of the burgeoning field of business sustainability within the forestry sector, this book paves a way forward in understanding what is working, what is not working, and what could potentially work to ensure sustainable business practices within the forestry sector.

Rajat Panwar is Assistant Professor in the Departments of Forest Resources Management and Wood Science in the Faculty of Forestry at the University of British Columbia, Vancouver, Canada. He is also Editor of the journal *Bioproducts Business*.

Robert Kozak is Professor and Head of the Department of Wood Science in the Faculty of Forestry at the University of British Columbia, Vancouver, Canada.

Eric Hansen is Professor of Forest Products Marketing in the Department of Wood Science and Engineering at the College of Forestry at Oregon State University, USA.

The Earthscan Forest Library

This series brings together a wide collection of volumes addressing diverse aspects of forests and forestry and draws on a range of disciplinary perspectives. Titles cover the full range of forest science and include the biology, ecology, biodiversity, restoration, management (including silviculture and timber production), geography and environment (including climate change), socioeconomics, anthropology, policy, law and governance. The series aims to demonstrate the important role of forests in nature, peoples' livelihoods and in contributing to broader sustainable development goals. It is aimed at undergraduate and postgraduate students, researchers, professionals, policy-makers and concerned members of civil society.

Recent Titles:

Forests, Business and Sustainability
Edited by Rajat Panwar, Robert Kozak and Eric Hansen

Climate Change Impacts on Tropical Forests in Central America: An Ecosystem Service Perspective
Edited by Aline Chiabai

Rainforest Tourism, Conservation and Management: Challenges for Sustainable Development
Edited by Bruce Prideaux

Large-scale Forest Restoration
David Lamb

Forests and Globalization: Challenges and Opportunities for Sustainable Development
Edited by William Nikolakis and John Innes

Smallholders, Forest Management and Rural Development in the Amazon
Benno Pokorny

Managing Forests as Complex Adaptive Systems: Building Resilience to the Challenge of Global Change
Edited by Christian Messier, Klaus J. Puettmann and K. David Coates

Evidence-based Conservation: Lessons from the Lower Mekong
Edited by Terry C.H. Sunderland, Jeffrey Sayer, Minh-Ha Hoang

Global Environmental Forest Policies: An International Comparison
Constance McDermott, Benjamin Cashore and Peter Kanowski

Monitoring Forest Biodiversity: Improving Conservation through Ecologically-Responsible Management
Toby Gardner, with a foreword by David Lindenmayer

Governing Africa's Forests in a Globalised World
Edited by Laura A. German, Alain Karsenty and Anne-Marie Tiani

Collaborative Governance of Tropical Landscapes
Edited by Carol J. Pierce Colfer and Jean-Laurent Pfund

Ecosystem Goods and Services from Plantation Forests
Edited by Jürgen Bauhus, Peter van der Meer and Markku Kanninen

Degraded Forests in Eastern Africa: Management and Restoration
Edited by Frans Bongers and Timm Tennigkeit

Forecasting Forest Futures: A Hybrid Modelling Approach to the Assessment of Sustainability of Forest Ecosystems and their Values
Hamish Kimins, Juan A. Blanco, Brad Seely, Clive Welham and Kim Scoullar

The Dry Forests and Woodlands of Africa: Managing for Products and Services
Edited by Emmanuel N. Chidumayo and Davison J. Gumbo

Forests for People: Community Rights and Forest Tenure Reform
Edited by Anne M. Larson, Deborah Barry, Ganga Ram Dahal and Carol J. Pierce Colfer

Logjam: Deforestation and the Crisis of Global Governance
David Humphreys, with a foreword by Jeffrey Sayer

The Decentralization of Forest Governance: Politics, Economics and the Fight for Control of Forests in Indonesian Borneo
Edited by Moira Moeliono, Eva Wollenberg and Godwin Limberg

Additional information on these and further titles can be found at http://www.routledge.com/books/series/ECTEFL

Forests, Business and Sustainability

Edited by
Rajat Panwar, Robert Kozak and Eric Hansen

First published 2016
by Routledge
2 Park Square, Milton Park, Abingdon, Oxon OX14 4RN

and by Routledge
711 Third Avenue, New York, NY 10017

First issued in paperback 2018

Routledge is an imprint of the Taylor & Francis Group, an informa business

British Library Cataloguing in Publication Data
A catalogue record for this book is available from the British Library

Library of Congress Cataloging in Publication Data
Forests, business and sustainability / edited by Rajat Panwar, Robert Kozak and Eric Hansen.
pages cm. -- (The Earthscan forest library)
Includes bibliographical references and index.
1. Sustainable forestry. 2. Forest products industry--Environmental aspects.
3. Sustainable development. I. Panwar, Rajat, editor. II. Kozak, Robert A., editor. III. Hansen, Eric, 1968- editor.
HD9750.5.F675 2016
338.1'7498--dc23
2015025144

Typeset in Goudy
by Taylor & Francis Books

ISBN 13: 978-1-138-58889-9 (pbk)
ISBN 13: 978-1-138-77929-7 (hbk)

Contents

Illustrations

Figures

Tables

Boxes

Contributors

Jim Bowyer, President, Bowyer & Associates, Inc. and Director, Responsible Materials Program, Dovetail Partners, Inc.

Benjamin Cashore, Professor, Environmental Governance and Political Science and Director, Governance, Environment and Markets (GEM) Initiative, Yale School of Forestry and Environmental Studies.

Alexandre Corriveau-Bourque, Project Manager – ICLA, Norwegian Refugee Council.

Peter deMarsh, Chair, International Family Forest Alliance and President, Canadian Federation of Woodlot Owners.

Chris Elliott, Executive Director, Climate and Land Use Alliance and Adjunct Professor, Department of Forest Resources Management, University of British Columbia.

Kathryn Fernholz, Executive Director, Dovetail Partners, Inc.

Reem Hajjar, Postdoctoral Research Fellow, International Forestry Resources and Institutions, University of Michigan.

Eric Hansen, Professor, Wood Science and Engineering Department, College of Forestry, Oregon State University.

Jani Holopainen, Doctoral student, Department of Forest Sciences, University of Helsinki.

Sébastien Jodoin, Assistant Professor, Faculty of Law, McGill University.

Peter Kanowski, Professor of Forestry and Master of University House, the Australian National University.

Robert Kozak, Professor and Head, Department of Wood Science, Faculty of Forestry, University of British Columbia.

Katja Lähtinen, Adjunct Professor, Department of Forest Sciences, University of Helsinki.

Duncan Macqueen, Principal Researcher – Forest Team, Natural Resources Group, International Institute for Environment and Development.

Constance L. McDermott, James Martin Senior Fellow, Environmental Change Institute, School of Geography and the Environment, University of Oxford.

Augusta Molnar (retired), Formerly Director, Country and Regional Programs, Rights and Resources Initiative.

D. Bryson Ogden, Private Sector Analyst, Rights and Resources Initiative.

Rajat Panwar, Assistant Professor, Departments of Forest Resources Management and Wood Science, Faculty of Forestry, University of British Columbia.

Romain Pirard, Senior Scientist, Centre for International Forestry Research.

Erica Pohnan, Conservation Program Manager, Yayasan Alam Sehat Lestari.

Anders Roos, Professor in Business Administration, Department of Forest Products, Swedish University of Agricultural Sciences.

Jacki Schirmer, Senior Research Fellow, Institute for Applied Ecology and Health Research Institute, University of Canberra.

Jenny Springer, Director, Global Programs, Rights and Resources Initiative.

Matti Stendahl, Researcher, Department of Forest Products, Swedish University of Agricultural Sciences.

Michael Stone, Doctoral candidate, Yale School of Forestry and Environmental Studies.

Anne Toppinen, Professor, Department of Forest Sciences, University of Helsinki.

Andy White, Coordinator, Rights and Resources Initiative.

Preface

It was in Venice some three years ago that we decided to write this book. Over pizza and beer by the storied canals, we discussed ways in which to reconcile the paradox of business and sustainability – the former fundamentally being about the extraction of resources and the latter fundamentally being about ensuring that resources remain for future generations. After studying (and teaching) business and sustainability in the forest sector for years, we shared a common frustration that there was no one consolidated source of information on this vast, multifaceted, nuanced, and perplexing topic. We were wary from the get-go of the daunting challenge of capturing the breadth of this topic in one book, yet we managed to identify a few core themes that spanned the bulk of extant sustainability literature. To be sure, this is not a review of business sustainability literature in the forest sector; yet, to be fair, it is an overview of the major directions in which this burgeoning field of study has grown (and continues to grow).

We would not have succeeded in bringing these vast strands of information together if it were not for the extraordinary support that was shown to us by our esteemed colleagues who have contributed to this book. In all earnestness, it is their book. Our thinking on business sustainability issues in the forest sector is profoundly shaped – and continues to be enriched – by these scholars, but also by our colleagues and the budding scholars in the Forests and Communities in Transition lab in the Faculty of Forestry at the University of British Columbia, and the Forest Business Solutions group in the College of Forestry at Oregon State University.

We hope that this book will help systemize learning at the nexus of business and sustainability in the forest sector, and will perhaps enable sustainability, business, and forestry scholars to refine their big-picture thinking on the various sustainability-based initiatives in the forest sector aimed at the same intended outcome – conserving the world's forests for the benefit of everyone on this planet.

Rajat Panwar
Robert Kozak
Eric Hansen

1 Many paths to sustainability, but where are we going?

Rajat Panwar, Eric Hansen and Robert Kozak

The scholar and social philosopher, James Carse, once remarked that "the mind does not come to life until it meets something it cannot comprehend." And that moment has arrived in the global forest sector.

Saving the world's forests is one of the grand challenges facing humankind (Howard-Grenville *et al.*, 2014). The timeframe within which to address this challenge before irreparable damage is done to the world's most diverse ecosystems is narrow. Ironically, those seeking to solve the problem are oftentimes the ones causing it. To make matters worse, there is no central authority dealing with the problem. Sustainability of the world's forests, then, by definition (Levin *et al.*, 2012), is truly a "wicked" problem.

While many philosophical approaches to address wicked problems seem enlightening, each remains a victim of its narrow paradigm – making stakeholder convergence a nightmare. Many solutions seem promising, yet each is restricted in scope and impact – making success a daydream. Baffled interest groups then begin to project the problem in their own ways, some by showcasing an inconvenient truth, others by constructing a convenient lie. As a result, targets become elusive, or worse even, illusive. Numerous initiatives – both substantive and symbolic – spring up from different directions. A wicked problem is thus accompanied by a multitude of proposed solutions, creating a vertigo-like situation.

It is against this backdrop that we view the numerous sustainability initiatives that permeate the forest sector. In the meantime, myriad forces are causing fundamental changes in the operational environments of the forest sector, as individual enterprises clamor to compete in the increasingly dizzying globalized markets of the twenty-first century. Reactions to these forces have resulted in the emergence of business models that not only must take sustainability into account, but must also allow for the pursuit of wealth creation through competitive advantage. We appear to be in the midst of a "sustainability vertigo," caught betwixt bewilderment that would puzzle even the most seasoned academics and exhilaration over the potential that forestry holds for affecting positive change.

Our primary motivation for this book is to help disentangle these perplexing issues and to help our readers to unpack the various strands of sustainability-oriented initiatives occurring in the forest sector. In so doing, this book draws on the expertise of globally renowned scholars that not only provide state-of–the-art knowledge on matters related to sustainability in the forest sector, but also critically assess the efficacy of a number of widespread sustainability initiatives.

The book is organized as follows. The book begins by sketching out a broad case for why we should care about the sustainability of global forests. In their chapter, "The spectrum of forest usage: From livelihood support to large scale commercialization," Kathryn Fernholz and Jim Bowyer take a functionalist view and argue, through data and anecdotes, that forests are indispensable to human well-being. They take a reader back to a time when forests were primary sources of food and fuel, and traverse into the modern era when – as we write this chapter – news comes of computer chips made of wood-based materials. The authors present, in a simple-to-understand manner, an excellent example of an institutional paradox (Thornton and Ocasio, 2008) that, on one hand, attributes human dependence on forest resources as a concern for sustainability, and, on the other, promotes increased uses of renewable forest resources as a means of addressing the sustainability challenge.

Paradoxes aside, Fernholz and Bowyer also provide an excellent backdrop for Constance McDermott to delve into the oft-ignored topic of our collective human footprint on forests in her chapter on regulatory and policy interventions for achieving sustainability. The issue of our human footprint has previously been raised on a global scale, most notably by Wackernagel and Rees (1996), and, more specifically, for the forest sector (Kozak, 2014). In both of these works, clarion calls for a check on consumption are made and the notion of a conservation-based economy is emphasized, yet the topic – for a variety of reasons – has not yet gained traction among sustainability scholars. In this light, McDermott's chapter is a persuasive reminder to the academic community that it is high time that the quest for sustainability confronts some difficult and inconvenient questions. This chapter presents a dense and comprehensive overview of the various governmental and stakeholder-driven policy initiatives, viewing them, on balance, as being fairly successful in helping to achieve sustainability.

This somewhat optimistic view of regulatory initiatives looks much more promising when juxtaposed with what Benjamin Cashore, Chris Elliott, Erica Pohnan, Michael Stone and Sébastien Jodoin have to say about the efficacy of market mechanisms in achieving sustainability: "[market mechanisms] ... seem poorly designed to fully address urgent issues such as the climate crisis and deforestation." For instance, forest certification was originally envisioned as a positive mechanism for sustainability of tropical forests, and in the past two decades, an enormous area of global forests has been third-party certified. Inevitably, the outside reviews of forest management practices have shown arguably incremental, management improvements. However, much of the world's tropical forests still remain largely outside the purview of forest certification despite the reassuring presence this can have in key global markets. Can forest certification as one emerging response then be described as a success? In their chapter entitled, "Achieving sustainability through market mechanisms," Cashore, Elliott, Pohnan, Stone and Jodoin explore these sorts of questions in more detail, focusing on the role of market forces in impacting forest sustainability. In so doing, they profile two key market forces in addition to forest certification, reduced emissions from deforestation and degradation (REDD+) and legality verification. The authors propose numerous ways in which market mechanisms could be strengthened to live up to their expectations, and by using four dimensions of market instruments and four pathways of influence, they describe conditions where market mechanisms can play a more prominent role in sustainable forestry.

The chapter on market mechanisms leads up to what has recently become the "holy grail" of sustainability initiatives across industries – corporate responsibility (CR). In their critical commentary, "On corporate responsibility," Anne Toppinen, Katja Lähtinen and Jani Holopainen first paint a broad theoretical canvas within which they situate ongoing CR practices in the forest sector. The chapter, broad in scope, rich in content, and balanced in judgement, calls for industry champions to catalyze forest sector companies in taking a leap – a quantum leap – from their entrenched conservative approaches and somewhat tainted past (where environmental issues are concerned) toward a resplendent future wherein they are proactive in addressing social and environmental concerns. In order for this future to unfold, the authors identify two major interventions: the development of a culture that is conducive to the implementation of CR practices and the integration of CR within broader strategic communications initiatives. In closing, the authors raise the issue of institutional factors required for making CR a success, perhaps reminding readers of the classic question of whether corporations can ever be responsible, or are they, by design, inherently wicked (Dunn, 1991)?

While it is beyond the scope of this book to address this question head-on, we do turn our attention to alternative forms of organization, land tenure reform and the emergence of new business models that many argue are central to the sustainability discourse in the forest sector. In the first of three chapters on this topic, Alexandre Corriveau-Bourque, Jenny Springer, Andy White and D. Bryson Ogden tackle the increasingly dynamic and highly political issue of forest tenure reform with an overview and treatise on the importance of recognizing community forest rights and the rights of Indigenous Peoples. The authors take a close look at how the ownership of the world's forests has changed in the past decade, and what implications this has had for local peoples, enterprises, governments, civil society and investors. Underlined by the fact that there, "is a growing base of evidence that demonstrates that local rights to land and forest resources are essential for the meaningful achievement of economic development, conservation, and climate change mitigation goals," the authors offer an account of how

Indigenous Peoples and other communities have successfully asserted their rights and increased their share of legal forest ownership of the world's forests, but caution readers that many claims to forests by local peoples remain contested and unresolved, with recent evidence suggesting a slowdown in the restitution of forest rights. In the final analysis, they put forward several logical and cogent policy-based arguments for local forest communities to continue down this decentralization path as a means of ensuring their sustainability and the sustainability of the forests upon which they depend.

Central to these questions on decentralization and forest tenure reform is the notion that these must go hand-in-hand with the creation of enabling environments which support the creation of viable small-scale enterprises and meaningful employment opportunities, especially in developing economies. The complexion of forest-based businesses is rapidly changing, and large multinationals – which have tended to dominate the forestry dialogue – are becoming increasingly inconsequential in a context of decentralization. In their chapter on "Enabling investment for locally controlled forestry," Duncan Macqueen and Peter deMarsh look at this question in more detail, from the point of view of investments in smaller-scale forestry operations, including community forest initiatives, smallholdings and small and medium enterprises. Opportunities for the creation of sustainable small-scale forest-based enterprises are virtually limitless, and flourishing examples abound from around the world. However, the success of locally controlled forestry is not without formidable challenges, not the least of which include a lack of market access to conventional markets, constrained access to capital, high transaction costs, and limited high-quality employment opportunities. The authors argue that investment is central to the success of locally controlled forestry and, ultimately, the long-term sustainability of forest resources; to that end, the chapter enumerates a number of practicable ways in which investment funds can be attracted. In addition, they offer up four recommendations on how enabling environments can be promulgated to increase the likelihood of success of locally controlled forestry and more meaningfully contribute to the livelihoods of forest-dependent communities.

In their chapter on "Decentralization and community-based approaches," Reem Hajjar and Augusta Molnar drill down even further on this topic with a summary of one of the most interesting and significant trends in the forest sector today. Decentralization – the process of national governments ceding powers to lower levels – is becoming increasingly commonplace in the forestry sector, typically manifested in the creation and management of community forests and community forest enterprises. While community-based approaches have had mixed results, they have certainly garnered much attention from civil society, policy makers and academics as a means of generating long-term economic, social and environmental benefits for vested communities. In this chapter, the authors – through a number of real-world examples – discuss the recent growth of decentralized models as alternative forms of tenure, provide a backdrop for why this trend is taking place, examine why it is that some community-based approaches succeed while others fail, and decisively show how such approaches can and do contribute to forest sector sustainability.

Up to this point, a recurring theme in this book is that "win–win" solutions are possible. A faith in this thesis – desirable as it may be – can only bring us so far. Management scholars are now beginning to see the naivety of a simplistic win–win view, and are instead focusing on the tensions that emerge on a path to sustainability (Hahn *et al.*, 2014). Jacki Schirmer, Romain Pirard and Peter Kanowski meticulously present those tensions in the context of plantation forestry in their aptly titled chapter, "Promises and perils of plantation forestry." Originally conceived to meet human fiber needs through planted forests and relieve pressure from pristine, ecologically valuable forests, plantation forests come with mixed baggage. Plantations currently provide about one-third of global industrial wood supply, yet that proportion remains much smaller than that supplied by natural forests. The authors present a cogent picture of the industrial forest plantation narrative and, in so doing, highlight the complexities inherent in expansionist interventions and "techno-fixes" to solve complex, multifaceted sustainability problems. This trend has done little to alleviate environmental concerns within society regarding forest sector activities. Instead, the concerns have simply shifted from issues such as clearcutting and the harvesting of old-growth to issues like monocultures, loss of biodiversity, water consumption, and, perhaps most critically, the impact of large plantations on local peoples. And yet, a genuinely sustainable business model remains an aspiration for the forest sector, and the plantation approach may mean improved competitiveness for many multinational forestry firms. With that paradox in mind, the authors call for a more sophisticated discussion of the merits of various models of plantation forestry and caution us about the danger of focusing only on the problems of the plantation approach.

And this is where our conversations specifically about sustainability of forests end, but a bigger interconnected question emerges: can forests play a role in global meta-level efforts toward planetary sustainability? Anders Roos and Matti Stendahl attempt to address this question within the futuristic concept of what has commonly been dubbed the "bio-economy." Climate change mitigation policies vis-à-vis the Kyoto Protocol and other policy instruments have driven extraordinary growth in the use of wood for energy, especially in European Union countries, not to mention about half of all fiber globally being used as energy sources for basic heat and cooking in less-developed countries. So, in some sense, these countries are ahead of the curve in thinking about the bio-economy compared to many developed economies. Beyond energy, there is a focus on wood as a source of chemicals via bio-refineries as a means of substituting petroleum-based products with renewables. Major governmental investments to prop up the renewable fuels and chemicals industries are now commonplace. Ironically, energy independence policies and recent crude oil price spikes have combined to create a glut of oil on global markets. This creates a major challenge for the emerging models of biomass for energy, fuel and chemicals. While Roos and Stendahl clearly see scope for these emergent bio-products, they also caution against a rising tension: while demand for renewable fiber may increase in the name of sustainable solutions, meeting that demand might lead to further sustainability threats.

And this is where the book ends. Never before has the need to reconcile business interests with the conservation of our forest resources – in a package that we like to call "sustainability" – been so pressing. These are complex, multifaceted issues that transcend a number of spatial and temporal scales, not to mention vastly differing stakeholder opinions on how the world should work. While we certainly don't have all of the answers, it is our hope that readers will appreciate the new thinking on forest sector sustainability presented in this book, and that it will provide some inspiration – or at least a renewed sense of purpose – in tackling this truly wicked problem.

References

Dunn, C.P. (1991). Are corporations inherently wicked? *Business Horizons*, 34(4), 3–8.

Hahn, T., Pinkse, J., Preuss, L., and Figge, F. (2014). Tensions in corporate sustainability: towards an integrative framework. *Journal of Business Ethics*, 127, 1–20.

Howard-Grenville, J., Buckle, S.J., Hoskins, B.J., and George, G. (2014). Climate change and management. *Academy of Management Journal*, 57, 615–623.

Kozak, R.A. (2014). What now, Mr. Jones? Some thoughts about today's forest sector and tomorrow's great leap forward. In *The Global Forest Sector: Changes, Practices, and Prospects*, E. Hansen, R. Panwar, and R. Vlosky (Eds). Boca Raton: CRC Press (Taylor & Francis Group).

Levin, K., Cashore, B., Bernstein, S., and Auld, G. (2012). Overcoming the tragedy of super wicked problems: constraining our future selves to ameliorate global climate change. *Policy Sciences*, 45(2), 123–152.

Thornton, P.H., and Ocasio, W. (2008). Institutional logics. In R. Greenwood, C. Oliver, R. Suddaby, and K. Sahlin-Andersson (eds), *The Sage Handbook of Organizational Institutionalism*, 99–128. London: Sage.

Wackernagel, M., and Rees, W. (1996). *Our Ecological Footprint: Reducing Human Impact on the Earth*. Gabriola, BC: New Society Publishers.

2 The spectrum of forest usage

From livelihood support to large scale commercialization

Kathryn Fernholz and Jim Bowyer

Forests play a key role in supporting the Earth's vital systems, and also in meeting a myriad of human needs. The fundamental purpose of this chapter is to capture the variety of ways in which forest resources have been and are used. We organize the chapter as follows: we first trace back in history how forest and forest products have grown in form and scope to meet various human needs; the subsequent sections address respectively the roles forests play in subsistence and developed economies.

Historical use of the forest and forest products

Humankind has for millennia depended on forests – primarily for wood – to meet various needs. Perhaps the first use of wood by people was for fuel. The first credible evidence of controlled use of fire was found in remains of campfires from about a million years ago – with fire likely used for cooking as well as heating (Miller 2013). In addition, wood was long ago used as part of simple tools and weapons. Wood also provided for shelter. A wooden framework draped with animal skins, dating back to 500,000–400,000 BC, and located within a cave, was found near Nice, France – the first known use of wood in this way. Evidence of freestanding wood huts, sometimes with low walls of stone and packed clay, has been found dating back to 450,000–380,000 BC. The use of wood frames, made of tree branches, to support animal skins, tree bark, and grasses continued through to at least 10,000 BC (Kowalski 1997, *Boundless* 2014).

Marked change in the way that wood was used did not occur until the development of tools that allowed cutting and shaping of logs and timbers. Early on, wood was shaped using a tool known as an adze. Consisting of a wooden handle made from a tree branch to which a blade was attached, the adze was pulled across wood surfaces to remove unwanted material. The first blades were made of hard stone, such as basalt or flint. Other early tools included axes and mallets/chisels, with both initially made of stone. Even these developments allowed great advances in wood use. Later, stone was replaced or tipped with copper, bronze, and then iron, each change allowing greater cutting ability. The advent of metals also created possibilities for development of more advanced tools. Woodworking tools included the axe, mallet and chisel, drill, file, and lathe. The first saws and files date back to the Bronze Age (700–500 BC), with the first lathes dating back to about 500 BC. Finished products were for the most part smoothed by rubbing with stones through the early part of the eighteenth century AD, although the first sandpaper made of crushed seashells bonded to parchment paper was used in China as early as the thirteenth century AD (Breasted 1906).

In ancient Egypt (about 3150 BC) substantial buildings were built, with adobe bricks and stone the primary materials used. However, considerable quantities of wood were also used in the form of pillars and lintels, and in roofs, doors, window shutters, and in fashioning upper stories of buildings. Counting empty beam holes and examining still-existing burned beams in building remains in Middle Bronze Age sites suggests wood use in excess of 2,000 trees per 150-room building. Wood was also extensively used for furniture, musical instruments, and everyday items of all kinds (Kuniholm 1997).

By 700 BC the first log houses had begun to appear in Europe, although it would be another 1,100 years before this type of construction became common, mainly in the abundantly forested countries of northern Europe. Shortly thereafter (about 600 AD) highly sophisticated construction of wood temples and shrines was being practiced in China and Japan; one existing temple, the Horyuji Temple in Japan, dates back to the year 607. The stave churches of Norway were built some 600 years later; several of these buildings, including the famous Heddal stave church that was built in the early thirteenth century, still exist today (Bramwell 1984).

By 1430 timber frame construction was well developed with wood structures six to seven stories in height built in the late Middle Ages. Spectacular hammer-beam roof structures topped churches and other public buildings in Europe, including London's Eltham Palace (1395) and Westminster Hall (1395–1399). Beyond structural applications, creativity on the part of woodworkers in using wood in a decorative way inside buildings flourished from the 1200s onward. The art of inlay or intasia – the laying of precious or exotic materials onto a base of solid wood – began in the Middle and Far East about 1400 BC, several hundred years after techniques for veneering of wood and gluing of veneers to form plywood were developed in Egypt. First used to create elaborate designs on small boxes and chests and on caskets and tombs of the rich and powerful, the art was later used in decorating walls, floors, and ceilings, reaching a peak in the middle and late eighteenth century. The art of creating mosaics, by joining hundreds or thousands of small wood pieces and shapes, also evolved from the tenth century onward. Timber frame construction was brought by early settlers from Europe to North America, and this type of construction was in evidence in the American colonies from 1636 onward. Balloon frame construction, similar to methods of wood house construction still in use today, became common with the advent of sawmilling and saw-milled planks about 1830 (Bramwell 1984).

As important as wood was (and is) in building construction, it played an even larger role in the development of transportation. Whereas previous to the fourth millennium BC simple tools and fire had been used in making rafts and dugout canoes to provide a means of water transportation, the ability to shape wood allowed the production of larger, more sophisticated vessels. Seagoing wood vessels were in use by about 3300 BC and shipments of timber from Lebanon to Egypt over the period 2686–2613 BC, and later from Crete, were referenced in ancient documents. Perlin (1991) and Oosthoek (2006) document how the fortunes of entire civilizations, including those of the Near East, Crete, Greece, Rome, and later Spain and Portugal, shifted as depletion of forest resources reduced sources of energy needed to smelt metals, and later to build ships and to thus participate in exploration and trade. Centers of economic wealth tended to move to regions where abundant forests remained.

The development of land transport was also almost totally dependent upon wood through the end of the nineteenth century. The first wooden wheels, fashioned from solid hewn wood or several planks fastened together with wood dowels, were used in Mesopotamia and in the locale of the present-day Netherlands as long as 5,000 years ago. Spoked wheels were in evidence by 2500 BC. And these wheels moved wooden carts, wagons, carriages, and even bicycles the world over through the early part of the twentieth century. The first automobiles had wood frames, as did airplanes through about 1920 and then again during World War II when wood use was revived in part to save steel for other uses (Bramwell 1984).

The development of tools that could be used to cut, shape, and drill wood was the factor most responsible for increasingly creative uses of wood through the middle of the nineteenth century. Thereafter, scientific inquiry played an increasingly important role.

An example of the role of science-based experimentation in expanding wood use is provided by the history of papermaking (Libby 1962). A precursor of paper, made of papyrus fiber arranged in crisscross fashion before pressing and drying, was in use in Egypt as early as 2400 BC. Use of parchment, made of skins of animals, followed. Paper was first made in China in about 105 AD from bamboo and the inner bark of mulberry in a process that involved soaking to soften fiber, pounding to reduce fiber to pulp, and then pouring into a frame fitted with a bottom of small strips of bamboo that allowed fiber to collect but water to drain away. Thereafter knowledge of paper-making spread slowly, and by 795 AD had spread to Samarkand (present-day northeastern Uzbekistan) where artisans soon learned to substitute fibers from linen rags for bamboo. Another 400–500 years would pass before knowledge of papermaking spread to Europe. Linen continued to be used, with straw eventually becoming the most common papermaking raw material as supplies of rags eventually proved insufficient. However, experimentation with wood in Germany in the mid-nineteenth century led to the invention of processes for pulping wood, first through mechanical means (1844) and then through the use of chemicals (1851). Further experimentation led to new methods of chemically pulping wood, with ground-breaking patents issued in the United States (1867), Germany (1884), and Sweden (1884). The first mill for producing what is currently known as kraft pulp was established in Sweden in 1890; kraft pulping remains the principal method for producing pulp worldwide today. Knowledge of how to use wood in making paper, coupled with mechanization of the papermaking process, served to markedly reduce the costs of paper and to facilitate communication worldwide; this development would also eventually translate to substantial demand on forests for pulpwood within the world's most economically developed nations.

The role of science-based discovery in changing the nature of wood use is also illustrated by early through recent history in the United States. Changing wood uses in the United States provide an example of early use based on relatively simple tools and techniques in cutting and shaping wood, and then marked change as tradition gave way to new products created through systematic research and development.

In colonial America (1607–1783), wood was the foundation on which society was built (Youngquist and Fleischer 1977). Buildings and furniture, spinning wheels and looms, dishes and pails, wagons and carriages, dinghies and ships, bridges and sidewalks, plows and hay rakes, milling machinery and sawmills, and products of every kind and shape were made of wood. Wood was also a major fuel source, used for heating and cooking and as the principal fuel of industry.

Wood use in the American Colonies, as in other parts of the world at that time, was not based on research, but rather on wood's abundance, range of inherent properties, ease of conversion to useful products, and long history of use in places of origin for America's immigrant population.

As the colonies gave way to rapidly expanding cities, and as populations expanded, wood abundance in many areas turned to scarcity as unrestrained wood use combined with land clearing for agriculture resulted in greatly diminished forests. But as wooden wagon trains carried homesteaders steadily westward, new forests were encountered and clearing of forests continued. Wood for fencing of pastures alone required enormous volumes of timber, with some 3.2 million miles of such fencing estimated to have been in existence in the mid-1800s. Development of the steam engine led to the need for great quantities of additional wood – for steamboat fuel and for railroad ties and trestles, which then offered a means of moving large volumes of wood to population centers. As in earlier times, research did not provide an underpinning for wood use.

One of the early drivers of inquiry into whether things might be done to increase the efficiency of wood use was the tendency of wood to rot. It was used in huge volumes for fencing, ties, trestles, bridges, and telegraph line poles, which required replacement after only a few years of use due to natural deterioration. As noted by MacCleery (1992), just replacing railroad ties on a sustained basis required from 15 to 20 million acres of forest land in 1900.

Interest in finding a way to preserve wood to eliminate or slow decay processes provided an impetus to an early field of inquiry which would later become known as the field of wood science.

As the population of the United States grew – from an estimated three million in 1785 to 77 million in 1900 – wood consumption grew rapidly; primary uses of wood were lumber and fuelwood. Rapid growth in wood use continued for about another decade but then, despite ongoing increases in population, a dramatic shift in wood use occurred.

First, lumber consumption declined almost as fast as it had increased (Figure 2.1). The causes of the decline were many, including substitution of non-wood materials such as steel and concrete for many applications, increased efficiency of wood use, and development of new technologies. Development of wood preservatives and preservative treatments alone resulted in a substantial reduction in the quantity of wood needed for replacement of ties, poles, fencing, and similar products. Another development, the invention of barbed wire, meant that as the 3.2 million miles of wooden fencing estimated to have existed in the mid-1800s began to deteriorate, far smaller quantities of wood were needed for replacement. In addition to declining lumber consumption, growth in the use of wood as a source of energy leveled off at the turn of the century and then began to decline as fossil fuels became increasingly important. Wood energy rebounded during the great depression of the 1930s, but then began a steep decline that continued through the early 1970s. By 1945, overall consumption of wood in the United States had fallen to a level similar to that of 1880 despite an almost three-fold increase in population during that period.

The lumber that was produced in North America had many uses, most of which were directly related to building and remodeling of houses. For example wood in the form of lumber was almost the only material used for framing of walls, floors, roofs, and siding in 1950. However, advances in plywood production arising from the work of aeronautical engineers and chemists during the course of World Wars I and II, had laid the groundwork for reliable construction-grade plywood; engineers tackled the challenge of

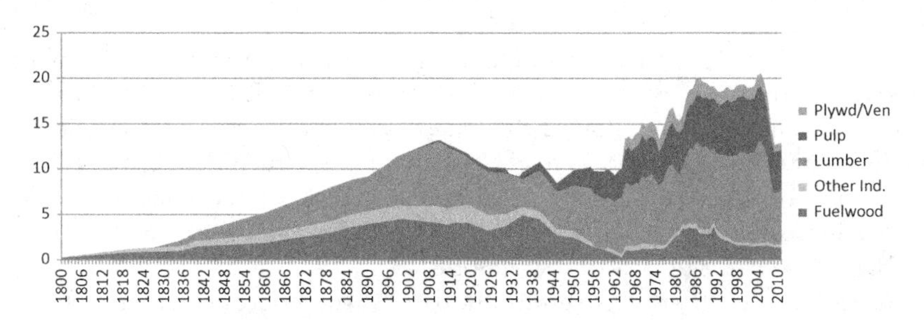

Figure 2.1 US consumption of wood and wood products, 1800–2011 (million cubic feet, roundwood equivalent)

Sources: 1800–1964, MacCleery (1992); 1965–2011, Howard and Westby (2013).

creating high-strength thin plywood while chemists played a critical role in developing waterproof glues needed for exterior use (Wood 1963). Another innovation involving gluing of wood, patented in Switzerland in 1897, was glue-laminated timber. A related patent in Germany (1906) was awarded for techniques to produce curved laminated timbers. Shortages of water-resistant casein glue during the war years delayed more widespread use of glued products, but these technologies were rapidly applied in building construction in the post-war years, a period characterized by a literal explosion of building activity, both in countries devastated by conflict and in those to which soldiers were returning (McNall and Fischetti 2014).

By 1960, the practice of using lumber for wall sheathing, subflooring, and roof decking had all but ended in North American construction, with all of these applications replaced by softwood plywood. Within another decade particleboard, a product made by compressing small particles or flakes of wood while simultaneously bonding them with an adhesive and that had been developed in post-war Germany, was in common use for subflooring and cabinet tops. A bonded fiber product – hardboard – was also often being used in place of plywood siding. Subsequent years would mark the introduction of oriented strand board (OSB), a product similar to but stronger than particleboard and made using thin-sliced, oriented wood strands; laminated veneer lumber (LVL), a "lumber" product made of parallel laminated veneers; parallel strand lumber (PSL), made using extrusion techniques; and laminated strand lumber (LSL), made in a similar fashion to OSB. Wood composite I-beams of high strength, but with smaller section modulus than solid beams, were also introduced during this period, as was plastic lumber for use in exterior decking. Fiber cement siding, made of wood fiber and cement, had been introduced as well and had captured much of the exterior siding market. All of these products are in common use today. Glue-laminated construction is also common in religious structures and increasingly employed in institutional and commercial construction (Bowyer *et al.* 2007).

While developments in the early twentieth century were inspired by wartime needs, later innovation was spurred by market competition. The new family of products made of fibers, particles, and flakes served to greatly expand options in wood products manufacturing and to increase the yield of final products. Structural composites such as LVL, PSL, LVL, and wood composite I-beams not only allowed the use of less wood for a given application, but also permitted production of large structural members from trees of small stature. At the same time, improvements in recycling technology greatly increased waste paper recovery and reuse rates, with these numbers up by 50–65% in the last 15 years alone.

As rapidly as wood consumption rose in the post-war years, the rise would have been far more spectacular were it not for innovation in both processes and products. For instance, in the twenty-five years between 1948 and 1973 the yield of lumber from a given quantity of logs processed in the

United States doubled, while the quantity of useful products obtained quadrupled (Bowyer *et al.* 2007).

Recent decades have brought a new stimulus for wood innovation. Concerns about energy efficiency of buildings and embodied carbon in building products have led to introduction of other new building products. One of these is the structural insulated panel, consisting of an insulating foam core sandwiched between two OSB panels. Another is cross-laminated timber (CLT), a thick panel product made of cross-oriented layers of lumber glued together in much the same way as veneers in plywood; these thick panels are used in what is described as mass timber construction, allowing creation of tall, energy-efficient, wood structures.

Energy and carbon concerns have also led to renewed interest worldwide in wood and other forms of biomass energy and in the possibility of producing current fossil-fuel-derived industrial chemicals from biomass sources. Wood has long been the principal source of energy for heating and cooking in the less-developed countries of the world. Presently, worldwide, about 53% of all wood consumed is used for home heating and cooking, with this consumption almost totally restricted to the world's developing economies (FAO 2013a). Energy use in the most economically developed countries has been dominated by fossil fuels over most of the past century, with little use of wood for heating and cooking. However, beginning in the late 1970s energy concerns resulted in efforts to recover energy from what had previously been manufacturing wastes. Most active in this regard has been the forest products industry itself. Material such as bark, sawdust, product trim and pulping liquors are now almost completely utilized, with much of the wood-derived energy produced by the forest products industry. Currently the primary wood products industries of the United States are about 60–70% energy self-sufficient, meaning that only 30–40% of the energy used for the manufacture of wood-based products is purchased from energy providers or produced from fossil fuels.

Very recently, governments and business enterprises around the world have focused on the low carbon emissions associated with biomass energy, and provided incentives for greater use of wood and other forms of biomass in energy production. Major initiatives are underway in a number of countries to increase production of biomass energy in the form of residential, institutional and district heating; electricity generation; and liquid fuels production. There is also considerable activity directed toward development of industrial chemicals from wood and other forms of biomass; chemical feedstocks used in production of plastics and a wide array of other products are currently largely obtained from fossil fuels.

Chemically, wood is composed of cellulose, hemicelluloses, and lignin. Cellulose and hemicelluloses, in turn, are made up of a number of five- and six-carbon sugars. There are many ways to transform these constituent sugars into useful products including a wide array of industrial chemicals such as lactic acid, ethanol, succinic acid, butanol, 1,4-butanediol (BDO),

3-hydroxypropionic acid, 1,3-propanediol, polyhydroxyalkanoates (PHAs), L-lysine, and more. All of these chemicals can be further converted into a number of derivatives. For instance, butanol is a platform chemical with several large volume derivatives, used as a solvent and in plasticizers, amino resins, and butylamines. Butanol can also be used as a bio-based transportation fuel and is more fuel efficient than ethanol on a volume basis in this application. As another example, PHAs are a family of natural polymers produced by many bacterial species. They are extremely versatile and can be used in a broad range of applications, including plastics (Bowyer and Ramaswamy 2007).

Another new and exciting use of wood biomass is in the developing field of nanotechnology. This focuses on materials at atomic and molecular scales. Potential applications range from use of nanofibers derived from plant materials as fundamental building blocks for new, lighter, and more energy-efficient materials that could conceivably substitute for plastic, metallic, or ceramic products; to coatings for enhanced performance; to components of new types of "intelligent" products that can alert users to potential problems.

These developments may signal the beginning of an entirely new era for wood and other biomass resources. Together, they have the potential to further increase the importance of wood to society and to greatly expand the value of forest sector markets.

Forest uses in subsistence societies and developing economies

Forests provide a great diversity of products and services. The United Nations estimates that 1.6 billion people – including 2,000 indigenous cultures – depend on forests for their livelihoods. Increasingly, research demonstrates that people and forests have co-existed and interacted as far back in history as science can take us. In parts of the world, the evidence for human interactions with the forest begins around the end of the last Ice Age. Pollen samples from tropical forest regions in Asia suggest people were clearing land, cultivating food, and planting imported seeds approximately 11,000 years ago (Garthwaite 2014). For example, in Borneo, pollen samples and charcoal evidence indicate a period of fire followed by the establishment of fruit trees, which is a departure from the plants that would have flourished after a wildfire and likely indicates land use change associated with human habitation and cultivation. In other regions, such as Australia and Africa, the evidence for human impacts on forests goes back 40,000–130,000 years (Mayell 2003).

The types of products and services that forests provide at any given time or in any given place are dependent upon both the capacity of the forest ecosystem as well as the needs of the people and communities that draw upon it. Forest uses are typically wide ranging. Priorities and dominant uses are influenced by the immediate needs and interests of the people that directly interact with the forest resources. In general, the products and services provided by forests in subsistence societies and developing economies include:

- wood for energy, heating and cooking
- shelter, other buildings, construction forms, bracing
- transportation conveyances including boats, carts and wagons, and boxes, crates, and other containers
- food resources, including from hunting, gathering, and production in agro-forestry systems such as sometimes used with coffee, bananas, and cacao
- medicines, traditional, scientific, and spiritual uses
- income from production and trade of forest products
- tourism, recreation, and aesthetic uses
- ecosystem services, carbon storage

These types of uses are briefly discussed in terms of their importance, impact, and trends in subsistence societies and developing economies.

Wood for energy, heating and cooking – a leading use of wood worldwide

The exploitation of forests as a source of energy is a leading use of wood worldwide. As noted earlier, about half of the world's annual harvest of wood is used as fuel (Table 2.1). The other half is categorized as industrial

Table 2.1 Global production of forest products, 2013

Product	*Million cubic meters*	*Million metric tons*
Roundwood	3,591	1,616[a]
• Wood fuel	1,854	834[a]
• Industrial roundwood	1,737	782[a]
Sawnwood	413	186[a]
Wood-based panels	358	161[a]
• Veneer and plywood	146	66[a]
• Particleboard and fiberboard	212	95[a]
Wood pulp		174
Other fiber pulp		14
Recovered paper		215
Paper and paperboard		398

Source: FAO (2013a).

[a] Conversion from cubic meters to metric tons based on assumption of an average green specific gravity of wood of 0.45.

roundwood to be used for various products and manufactured materials. An estimated 3.5 million cubic meters of wood are harvested annually, and about 1.8 million cubic meters are used as wood fuel for energy. The use of wood for heating and cooking is a traditional and historic use of forest materials. In developing economies, there is increasing attention being paid to providing sustainable sources of fuelwood. There are also initiatives to reduce the negative human health impacts of indoor combustion and increase the efficiency of the fuel-to-energy conversion. Wood fuel production is of leading importance in Africa, where fuel accounts for 90% of roundwood production. In the Asia-Pacific region, wood fuel is 65% of roundwood production; and in Latin America and the Caribbean it is estimated at 54% (FAO 2010). In contrast, in Europe and North America, wood use for energy production represents a relatively minor portion of roundwood production, at 20% and 9%, respectively. However, in the period 2009–2013, energy wood production in Europe increased by 20% perhaps signaling a significant change in wood use (FAO 2013a). Also as noted earlier, in developed nations, including many parts of the European Union, wood energy is increasingly being utilized as a renewable energy alternative to fossil fuel resources. Wood is the leading source of renewable energy in many regions. Wood fuel and charcoal continue to be important wood energy sources globally, and the expansion of wood pellet production has created new opportunities for wood fuel use and global trade. Wood energy can help governments meet policy goals for reduced carbon emissions. Advances in thermal wood energy technologies and biofuels indicate that the utilizing wood for energy is likely to continue to be a primary use of wood throughout the world for many years to come.

Shelter – uses of wood for structures

Wood is a primary building material worldwide. From logs to lumber, panels, and engineered wood products – many different forms of wood are used to construct homes and a full range of building types. Wood is an important structural material. It is also used for non-structural building components such as roofing, siding, and finishing products. In developing regions, wood provides an important resource for constructing shelter because wood is frequently locally available and affordable. Constructing a home from wood is also a relatively light form of construction that does not require heavy and expensive equipment or tools. In those countries characterized by abundant forest resources and well-developed economies, wood is the principal material used in constructing homes. Such countries include the United States, Canada, and those of Oceania (Australia and New Zealand), and Northern Europe. Considerable quantities are used for this purpose in Japan as well, where about one-half of homes are primarily wood. In Central and Eastern Europe wood is used mostly in windows, doors, and roofs, and interior finishes of houses, although wood-frame construction is gaining in popularity in some countries. Wood and/or bamboo are also the materials of choice for house building in rural areas of many developing countries blessed with significant forest cover. Beyond housing, a recent trend in the most economically developed countries has been to look for new uses of wood as a building material, including initiatives to support commercial and non-residential construction using wood because of the environmental benefits.

Transport – boats, railroads, wagons/carts, crates, and bridges

A number of attributes of wood have made it a useful material for transportation and packaging. Though no longer the dominant material for shipbuilding and production of overland vehicles as in earlier times, wood remains important in transportation in the form of packaging and shipping containers, pallets, and skids. The trailers of large semi-trailer trucks and transports are commonly made primarily of plywood, as are box car liners and railroad ties throughout North America. Bridges have also been an important use for wood in transportation systems. In developing economies, wood is often a locally available, affordable, and durable material to use in transportation systems. Several universities around the world have active programs addressing timber bridges and wood transportation structures, including in Portugal, Brazil, Chile, Spain, and the United States (National Center for Wood Transportation Structures 2015).

Forests as a source of food

Forest ecosystems are an important source of food. Wildlife from forests can be essential sources of food in subsistence and developing economies. Researchers have estimated that in 62 developing countries, people rely on wild meat and fish for at least 20% of the protein in their diets. However, this use can be harmful to wildlife populations if not appropriately managed. Forest-based communities may also utilize many other materials from forests and other habitats as food sources. Estimates are that there are more than 1,900 edible insect species and that as many as two billion people around the world include a variety of insects in their regular diets (FAO 2013b). Forests also provide fruits, nuts, and berries. Diverse tree species provide products such as syrup or sugar, chewing gum, edible needles and leaves, and tender inner bark. Forest productivity can be enhanced through forest farming or agroforestry systems that are well established in some cultures. Forests may also be used for rotational grazing or shifting agriculture. These land-use dynamics can result in major or complete loss of forest-derived benefits when economics or preferences dictate permanent or semi-permanent conversion to non-forest uses. Conversions to intensive annual crop production, oil palm, or rubber plantations reduce the biodiversity, ecosystem resilience, and co-benefits and are examples of land use shifts that result in major loss of forest services (UNESCO 2010).

Forests as a source of medicines and scientific discoveries

Forests continue to be a source of scientific breakthroughs and discoveries. New plants and animal species are continually being discovered around the world. In 2014, more than 200 new species were discovered, including 110 ants, 16 beetles, three spiders, 28 fish, 25 plants, and one mammal (Griffiths 2014). Scientists are also gaining insights into human history through forest-based discoveries. Forests have also provided significant medical advances and are important to traditional and spiritual values in diverse cultures. Researchers have estimated that in India about 2,500 plants have medicinal uses and at least 400 million Indians rely upon traditional medicine. In Africa, 70–80% of the population may rely on medicinal plants. In the United States, it is estimated that annual sales of pharmaceuticals of natural origin exceed US$75 billion (FAO 2006).

Forests as a source of rural income

There are many ways in which forests can provide a source of income. Forest materials, such as roundwood (logs) or other wood products are harvested, manufactured into value-added products, and sold to local or export markets. In forested regions the availability of wood-product markets can be an important part of the economy. However, in some instances, illegal or exploitative logging can contribute to forest degradation as well as social injustice. Verifiable land tenure and ownership and/or better laws and enforcement are often needed before sustainable management programs can be established (Illegal Logging Portal 2015).

A wide range of non-timber forest products may also be gathered and marketed as sources of income. These product markets may include foods, medicines, essential oils, and other materials. In addition to timber and non-timber forest products, forests can provide other diverse income and economic development opportunities, for example, through forest-based travel and eco-tourism.

In recent years, efforts have been made to provide payments related to the ecosystem services that forests provide, including payments for carbon storage, water quality protection, wildlife habitat, and other benefits. These efforts have primarily been focused on forested tropical countries in order to reduce economic pressures for conversion of diverse forest ecosystems to other land uses such as agriculture. A concept known as REDD (Reducing Emissions from Deforestation and forest Degradation) is guiding efforts in developing countries, and a number of initiatives are underway. Since initial discussions in 2005, the concept has evolved into REDD-plus or REDD+, which refers to "Reducing emissions from deforestation and forest degradation and the role of conservation, sustainable management of forests and enhancement of forest carbon stocks in developing countries." Global and multi-national initiatives currently supporting REDD+, include the United Nations

Collaborative Programme on Reducing Emissions from Deforestation and Forest Degradation in Developing Countries (UN-REDD Programme), the Forest Carbon Partnership Facility (FCPF), and the Forest Investment Program (FIP) hosted by the World Bank. The intent is to create an incentive for developing countries to protect, better manage, and wisely use their forest resources, and in so doing contribute more effectively to conserving biodiversity and addressing climate change (UN-REDD 2015). In addition to environmental benefits, REDD+ also offers social and economic benefits and it is being integrated into green economy strategies. The UN-REDD Programme provides technical and financial support to developing countries to establish the capacity necessary to implement REDD+. The implementation of REDD+ activities can be challenging, in part due to capacity and governance issues. There are also factors such as agriculture, food security, and rural economies to consider. To date, much of the work has focused on development of systems to support REDD+ activities. Now it is recognized that more progress is needed in implementation. As of January 2015, five countries (Brazil, Colombia, Guyana, Malaysia, and Mexico) have submitted reference emission levels for assessments following the REDD+ process (UN-REDD 2015). More on REDD+ will follow in subsequent chapters of this book.

Forest uses in developed economies

Differences in use of forest resources by geographic region are shown in Table 2.2, including significant differences between regions in the balance between use of harvested forest resources for wood fuel vs. industrial roundwood. In the economically most developed countries with large forest estates, the vast majority of wood harvested is used as industrial roundwood; in these regions, wood and other products of the forest often support diverse industries that contribute significantly to regional and national economies.

The use of forests in developed countries includes a full range of traditional as well as innovative products, including:

- Building products, lumber and sheathing, fencing, railroad ties, utility poles, landscaping products, and other solid-wood forest products.
- Paper, packaging, sanitary and medical products, and other fiber materials.
- Energy, including thermal home heating, institutional, and district heating; and electric supplies or co-generation facilities.
- Bio-energy, including pellet fuels for domestic use and export, liquid biofuels, and other innovations.
- Bio-chemicals, wood-derived flavorings and additives.
- Food, medicine, ornamentals – timber and non-timber forest products.

Many of these products and their evolution were discussed previously in the section looking at historic forest uses. A few key points to consider are

Table 2.2 Regional wood removals, 2005 (million m^3)

Region/subregion	*Industrial roundwood*	*Woodfuel*	*Total*
Eastern and Southern Africa	35	174	209
Northern Africa	4	24	28
Western and Central Africa	24	297	321
Total Africa	63	495	558
East Asia	86	71	157
South and Southeast Asia	99	463	562
Western and Central Asia	17	13	30
Total Asia	201	547	748
Total Europe	560	164	724
Caribbean	1	5	6
Central America	3	16	19
North America	701	55	756
Total North and Central America	705	76	781
Total Oceania	55	1	56
Total South America	178	167	345
World	1762	1449	3211

Source: FAO (2010)

that research shows that economies that develop and continue a reliance on forest resources are likely to continue to retain forest land. In contrast, countries or regions with economies that do not develop a forest-based economy are more likely to see forested areas converted to other land uses, such as livestock or annual crop production. Also, as has been described previously, the ways in which a country, community, or culture utilizes its forests frequently changes over time as needs, resources, technologies, and other factors also change.

Recent decades have brought a new stimulus for wood innovation, especially in developed economies. Concerns about energy efficiency and carbon impacts of construction have led to new wood building products and innovations in design. Energy and carbon concerns have also led to renewed interest in wood as a source of renewable biomass energy. As discussed in the section on the history of wood use, a number of current initiatives globally are focused on greater use of wood and other forms of biomass for energy production. In this regard, many governments are providing incentives for biomass energy. Some countries are also promoting development of wood- and biomass-based industrial chemicals.

Also noted earlier is the developing field of nanotechnology that focuses on materials at atomic and molecular scales. Wood is an attractive potential source of nanofiber, a reality that could create new markets for forest resources and connections to emerging high-value industries.

Forest services

Forests provide a full range of products and services. Sometimes there is a focus on the tangible products that forests provide, such as lumber, fiber, and food, due in part to their clear economic importance. In the category of forest services, there are other benefits derived from forests that further enhance quality of life, human health, and ecosystem functions. Some of these benefits and services have been monetized or could be monetized in the future. Examples of forest services include:

- recreation
- wildlife habitat
- watershed and water supply protection
- aesthetics, scenic beauty, and tourism
- ecosystem services, including carbon storage

These examples of forest services are briefly described including their economic impacts and current trends.

Recreation

Forest-based recreation includes hiking, camping, hunting, wildlife watching, and other motorized and non-motorized activities. Recreation in many parts of the world commonly occurs on publicly-owned lands, including national, state/provincial, or local parks or forests. Recreation also occurs on private lands through formal or informal arrangements. For example, in Nordic countries, private lands remain open for certain types of public recreational use, such as hiking, skiing, and short-term camping (VisitNorway.com 2015). In the United States, it is common for the recreational use of private lands to occur through a lease agreement or other formal arrangements between the landowner and user. Inappropriate or inadequate management of recreational uses can have negative impacts on forests, including the spread of invasive species; soil compaction, erosion, or other degradation; wildlife disruption (e.g. disturbing birds during the nesting season); or other impacts. Responsible recreation management is increasingly important in areas that are experiencing growing pressures due to expanding populations (e.g. recreational areas near urban centers) or due to diversifying and competing recreational interests that may not be compatible. According to a 2012 report, outdoor recreation in the United States provides more than six million jobs and includes US$646 billion in direct consumer spending annually (Outdoor Industry Association 2012). Developing economies, in general, may have limited forest recreation opportunities, with those that do exist largely dedicated to tourism.

Wildlife habitat

Forests provide an important habitat for a full range of wildlife species throughout the world. Common species as well as rare or endangered species benefit from diverse and abundant forest habitats. It is widely reported that some 80% of the Earth's biodiversity depends upon healthy forest ecosystems. Management of forests to meet specific wildlife habitat needs can be important for addressing habitat limits or threats to wildlife populations. For example – Kirtland's warbler (*Setophaga kirtlandii*), an endangered migratory songbird species in the Midwestern United States, requires a very specific type of fire-dependent forest consisting of young (5–23 year old) jack pine trees mixed with grass openings where they build their ground nests. In the absence of fire, forest management practices and well-designed harvesting treatments have been important tools for restoring Kirtland's warbler populations in the state of Michigan (USFWS 2015a). Similarly, the Karner Blue butterfly (*Lycaeides melissa samuelis*) is an endangered species found in the Northeastern United States with specific habitat needs for pine and scrub oak forests mixed with grassy areas and their food plant the wild lupine (*Lupinus perennis*). Active forest management and habitat restoration has been important for protecting populations of this butterfly (USFWS 2015b). The role of forests and forest management continues to be debated for a variety of wildlife species. In many instances, there are multiple factors impacting the health of a specific species. For example, the Northern spotted owl (*Strix occidentalis caurina*) was impacted by habitat loss; however, biologists have also determined that invasion by the more aggressive barred owl has also impacted breeding success rates for the spotted owl and is a significantly greater threat to spotted owl recovery than was initially recognized (USFWS 2011).

Watershed protection, municipal water supplies

Forests and water resources interact in many ways. Forests are important for preventing soil erosion and protecting water quality. A study of the northeastern United States found that 50 to 75% of the region's population relies on surface waters for their drinking water, affecting more than 52 million people from nearly 1,600 communities (Barnes *et al.* 2009). Surface waters from forested watersheds consistently provide some of the highest-quality drinking water in the world. Many communities have established programs to protect forested watershed and support forest land uses that maintain high water quality, including the cities of New York, Boston, and Denver.

Scenic beauty, tourism

In many places, forests are the scenic backdrop to people's lives. Rolling green forest-covered hills and mountains, a flush of spring flowers, or the multi-colored palette of deciduous trees in autumn mark the change of the seasons. The economic value of the scenic beauty of forests can be estimated by the associated tourism. For example, in New England, there are Fall Color tours that include cruise ships, buses, and self-guided driving tours and hikes. Restaurants and lodgings cater to visitors from around the world that come to experience the beauty. In 2014, it was estimated that tourists would spend at least US$3 billion to view the autumn beauty of New England. The State of New Hampshire projected 8.3 million visitors in 2014, up from 7.5 million in 2009 (Associated Press 2014).

Carbon storage, ecosystem services

Since the 1990s, there has been increasing policy attention given to the ecosystem services that different land uses provide, including the carbon storage benefits of forests. Forests are recognized to be a significant factor in the carbon cycle. Globally, forests are estimated to contain 80% of above-ground and 40% of below-ground terrestrial carbon (UNFCC 2015). Also, forest products continue to store carbon as long as they exist. Wood is one-half carbon (by weight), meaning that wood homes and other structures, furniture, and other wood products store substantial quantities for carbon for the long term. Providing incentives to grow and care for forests has been an important part of climate change policies and mitigation efforts in many parts of the world. In the United States, state-level programs include recognition of forest carbon offset projects. Globally, the REDD initiatives support investments in forestry in developing nations to help meet carbon goals and reduce the threat of forest land conversion to non-forest uses, such as intensive agriculture which has reduced carbon storage benefits.

Conclusion

Humans have relied on forests for many thousands of years, as sources of food, fuel, and shelter. Products made from wood have become increasingly sophisticated over the past century as research and development efforts have improved and innovated.

Whereas utilization of wood as a fuel for heating and cooking has been a dominant use for millennia, the value of wood as an energy source is receiving renewed attention. Initiatives have been mounted worldwide to increase use of wood in energy generation, including electricity, thermal energy, and liquid fuels. A wide range of industrial chemicals and nanomaterials are also within the realm of possible new forest products. At the same time that increased emphasis is being given to the tangible products of the forests, there is greater awareness of amenity values. The need for watershed protection has never been greater. The carbon capture and storage potential of forests and forest products is in the spotlight. Biodiversity is of great concern. And, recreation and aesthetics are high on the list of coveted forest values.

Retaining forests and ensuring their responsible management will continue to be a challenge in the decades and centuries ahead. Rising populations and increasing demands on forests can be expected to heighten concerns about forest sustainability. On the one hand, increased use can increase value and thereby increase incentives to retain forests. On the other hand, high value in the absence of adequate laws and governance can increase risks of irresponsible exploitation. Careful thought and planning will be needed to ensure abundant and healthy forests for future generations. The global patterns of forest usage reflect the needs of people and communities. These patterns and the needs they represent are the foundation for making a connection between forest usage and forest sustainability, as discussed in subsequent chapters.

References

Associated Press. 2014. "New England's fall foliage draws economic boon of leaf peepers." October 19. (www.tampabay.com/features/travel/new-englands-fall-foliage-draws-economic-boon-of-leaf-peepers-wvideo/2202854) Accessed 6 Jan. 2015.

Barnes, M., Todd, A., Whitney Lilja, R., and Barten, P. 2009. *Forests, Water and People: Drinking Water Supply and Forest Lands in the Northeast and Midwest United States.* USDA Forest Service, June. (http://na.fs.fed.us/pubs/misc/watersupply/forests_water_people_watersupply.pdf)

Boundless. 2014. Paleolithic architecture: shelter or art? *Boundless Art History.* July 3. (www.boundless.com/art-history/textbooks/boundless-art-history-textbook/prehistoric-art-2/the-paleolithic-period-45/paleolithic-architecture-shelter-or-art-271-5308) Accessed 10 Jan. 2015.

Bowyer, J. and Ramaswamy, S. 2007. *An Assessment of the Potential for Bioenergy and Biochemicals Production from Forest-Derived Biomass in Minnesota.* Dovetail Partners, Inc., August 29. (www.dovetailinc.org/report_pdfs/2007/blandinirrbioenergypaper082907yf.pdf)

Bowyer, J., Shmulsky, R., and Haygreen, J. 2007. *Wood in the Global Raw Materials Picture. Forest Products & Wood Science*, 5th Edition, Chapter 19. Oxford: Wiley-Blackwell.

Bramwell, M. 1984. *The International Book of Wood.* New York: Crescent Books.

Breasted, J. 1906. *Ancient Records of Egypt, Part One: Historical Documents from the Earliest Times to the Persian Conquest.* Chicago: University of Chicago Press (as reported by reshafim.org). (www.reshafim.org.il/ad/egypt/timelines/topics/wood.htm) Accessed 5 Jan. 2015.

FAO (Food and Agriculture Organization of the United Nations). 2006. Forests and human health. *Unasylva* 57(224). (ftp://ftp.fao.org/docrep/fao/009/a0789e/a0789e.pdf)

FAO (Food and Agriculture Organization of the United Nations). 2010. *Global Forest Resources Assessment.* (www.fao.org/forestry/fra/fra2010/en)

FAO (Food and Agriculture Organization of the United Nations). 2013a. *Forest Product Statistics. Global Forest Products Facts and Figures.* (www.fao.org/forestry/statistics/80570/en)

FAO (Food and Agriculture Organization of the United Nations). 2013b. *Edible Insects: Future Prospects for Food and Feed Security.* (www.fao.org/docrep/018/i3253e/i3253e.pdf)

Garthwaite, J. 2014. "The remnants of prehistoric plant pollen reveal that humans shaped forests 11,000 years ago." Smithsonian.com, March 5. (www.smithsonianmag.com/science-nature/remnants-prehistoric-plant-pollen-reveal-humans-shaped-forests-11000-years-ago-180949985/?no-ist) Accessed 5 Jan. 2015.

Griffiths, S. 2014. "Elephant shrew and sea slugs among 221 new species of animal discovered this year." DailyMail.com, 30 December. (www.dailymail.co.uk/sciencetech/article-2891121/Elephant-shrew-sea-slugs-221-new-species-animal-discovered-year.html) Accessed 5 Jan. 2015.

Howard, J. and Westby, R. 2013. *U.S. Timber Production, Trade, Consumption and Price Statistics 1965–2011*. Research Paper FPL-RP-676. U.S.D.A. Forest Service, Forest Products Laboratory. (www.fpl.fs.fed.us/documnts/fplrp/fpl_rp676.pdf)

Illegal Logging Portal. 2015. Chatham House. (www.illegal-logging.info/topics/causes) Accessed 6 Jan. 2015.

Kowalski, W. 1997. "Stone age habitats." *Aerobiological Engineering*. (www.aerobiologicalengineering.com/wxk116/StoneAge/Habitats) Accessed 10 Jan. 2015.

Kuniholm, P. 1997. Wood. In: Meyers, E. (ed.) *The Oxford Encyclopedia of Archaeology in the Near East*, New York: Oxford University Press, pp. 347–349. (http://dendro.cornell.edu/articles/kuniholm1997b.pdf) Accessed 7 Jan. 2015.

Libby, E. 1962. History of Pulp and Paper. In: Libby, E. (ed.) *Paper Science and Technology, Vol. 1*. New York: McGraw-Hill Book Company, pp. 1–19.

MacCleery, D. 1992. *American Forests: a History of Resiliency and Recovery*. Forest History Society. (www.foresthistory.org/publications/Issues/American_Forests.pdf)

Mayell, H. 2003. "First humans in Australia dated to 50,000 years ago." *National Geographic News*, 24 February. (http://news.nationalgeographic.com/news/2003/02/0224_030224_mungoman.html) Accessed 5 Jan. 2015.

McNall, A. and Fischetti, D. 2014. Glue laminated timber. In: Ball, T. (ed.) *Twentieth Century Building Materials – History and Conservation*. The Getty Conservation Institute. Los Angeles: Getty Publishers, p. 105.

Miller, K. 2013. "Archaeologists find earliest evidence of humans cooking with fire." *Discover Magazine*, May 9. (http://discovermagazine.com/2013/may/09-archaeologists-find-earliest-evidence-of-humans-cooking-with-fire) Accessed 8 Jan. 2015.

National Center for Wood Transportation Structures. 2015. Iowa State University, Institute for Transportation. (http://woodcenter.org) Accessed 5 Jan. 2015.

Oosthoek, K. 2006. "The role of wood in world history." *Environmental History Resources*. (www.eh-resources.org/wood.html) Accessed 8 Jan. 2015.

Outdoor Industry Association. 2012. *The Outdoor Recreation Economy 2012*. (http://outdoorindustry.org/pdf/OIA-RecreationEconomyReport2012-TechnicalReport.pdf) Accessed 5 Jan. 2015.

Perlin, J. 1991. *A Forest Journey: The Role of Wood in the Development of Civilization*. Cambridge, MA: Harvard University Press. (www.sciencedaily.com/releases/2014/01/140124082608.htm). Accessed 5 Jan. 2015.

UNESCO. 2010. *Protecting Biodiversity, Protecting Forests*. (www.unesco.org/new/en/media-services/single-view/news/protecting_biodiversity_protecting_forests/back/18256/#.VKwhZqYwYVE) Accessed 6 Jan. 2015.

UNFCC. 2015. Submissions from Parties on Proposed Forest Reference Emission Levels and/or Forest Reference Levels for the Implementation of the Activities Referred to in Decision 1/CP.16, paragraph 70. (http://unfccc.int/methods/redd/items/8414.php) Accessed 13 Jan. 2015.

UN-REDD Programme. 2015. (www.un-redd.org) Accessed 6 Jan. 2015.

USFWS (US Fish & Wildlife Service). 2011. "Revised recovery plan for the Northern Spotted Owl (*Strix occidentalis caurina*)." Region 1. Portland, Oregon: U.S. Fish and Wildlife Service. (http://ecos.fws.gov/docs/recovery_plan/RevisedNSORecPlan2011_1.pdf)

USFWS (US Fish & Wildlife Service). 2015a. Environ. Conserv. Online System. "Species profile for Kirtland's warbler (*Setophaga kirtlandii*)." (http://ecos.fws.gov/speciesProfile/profile/speciesProfile.action?spcode=B03I) Accessed 6 Jan. 2015.

USFWS (US Fish & Wildlife Service). 2015b. Environ. Conserv. Online System. "Species profile for Karner Blue butterfly (*Lycaeides melissa samuelis*)." (http://ecos.fws.gov/speciesProfile/profile/speciesProfile.action?spcode=I00F) Accessed 6 Jan. 2015.

VisitNorway.com. 2015. *Right of Access*. (www.visitnorway.com/us/about-norway/travel-facts/when-you-arrive/right-of-access) Accessed 6 Jan. 2015.

Wood, A. 1963. *Plywoods of the World: Their Development, Manufacture, and Application*. Edinburgh: Morrison and Gibb, Chapter 1.

Youngquist, W. and Fleischer, H. 1977. *Wood in American Life 1776–2076*. Madison, WI: Forest Products Research Society.

3 The regulatory and policy interventions for achieving sustainability

Constance L. McDermott

This chapter provides a broad historical overview of state-based domestic and international forest-related policies. The overview is guided by two different theoretical perspectives – referred to as "forest transitions" and "footprints", respectively – that explain why policies have evolved the way they have, and what the implications are for effective forest governance into the future.

Governments have long played a pivotal role in steering forest change. However, the goals and strategies shaping this role continue to change over time, and are strongly linked to the broader dynamics of a country's social, economic and political development. Just what the nature of that link is, and how society can best balance economic development with forest conservation, is a matter of considerable debate. On the one hand, research on "forest transitions" has shown how higher levels of economic development have been historically correlated with better forest conservation within a country's borders. On the other hand, research on global environmental "footprints" suggests that global trade has enabled developed countries to consume more resources per capita than the earth can sustainably provide. These higher rates of consumption have been made possible by the displacement of environmental harm and deforestation to the world's less-developed regions. Thus some researchers question whether the world's forests can be effectively conserved if consumption rates continue to rise.

The forest transition and footprint perspectives, in turn, suggest different ways to understand and evaluate government forest policies. Rather than take sides in the debate, this chapter draws on both viewpoints in order to understand what is driving today's key forest challenges and to uncover possible solutions. It begins by elaborating a little further on the research informing the forest transitions and footprint perspectives. It then considers what these perspectives tell us about trends in domestic forest policies. This is followed by an analysis of state-based efforts to achieve global sustainability through international forest governance. Finally, I conclude by assessing all of these trends in light of both the "forest transitions" and "footprints" perspectives, and exploring ways forward.

Forest transitions

There is a growing body of literature that has examined the drivers of forest change. This research has highlighted how governments in less-developed countries have frequently enacted laws and policies that encourage the conversion of natural forests to other uses. For example, states have sponsored programs to resettle urban or rural populations within the forest frontier and established tenure regimes that grant land rights to farmers who clear the forest (e.g. Rudel *et al.*, 2009). To the extent that forest protection is a concern, less-developed states often lack the economic resources and/or political commitment to achieve it. However, as deforestation and development advance, countries eventually undergo a "forest transition" characterized by a slowing and reversal of forest loss (Mather, 1992). Explanations for this forest transition are many, and vary by country. But core drivers include shifts from resource-based to manufacturing- or service-based economies, increasing urbanization and farm abandonment, growing wood scarcity and/or increasing intensity of wood production, rising wood imports, and shifts in societal values (Barbier *et al.*, 2010). At the global scale, this "forest transition" has meant that deforestation of the global North had already peaked by the turn of the twentieth century and most Northern countries are currently gaining forest cover. Meanwhile, deforestation in the global South has been accelerating since the mid twentieth century (FAO, 2013).

Not only is there variation in the causes of forest transitions, but the role and effect of forest policy on these transitions also appears to vary. A comparison of forest policies in 20 countries and 45 jurisdictions worldwide shows considerable variation among both developed and developing countries in regards to regulatory approaches and requirements (McDermott *et al.*, 2010). I provide a brief review of the findings here, as a means to summarize the range of approaches currently pursued by governments at the domestic level, as well as to compare and contrast policy dynamics in developed (post-transition) and developing (pre-transition) countries.

The McDermott *et al.* (2010) study addressed a range of key forest practice and landscape-level environmental issues, and classified them according to a policy typology. The typology firstly considered whether policies existed to address the key issues identified, and if so whether they entailed voluntary guidelines (frequently referred to as Best Management Practices) or mandatory requirements. It also distinguished between "procedural" policies, which address forest planning and/or procedures for decision-making, and "substantive" policies that entail standardized qualitative or quantitative threshold requirements for environmental protection. Finally, it identified "results-based" approaches that aimed to prescribe desired outcomes, e.g. particular levels of water quality, rather than dictate particular practices, e.g. no timber harvest within 50 meters of the stream edge. Policies were then ranked according to their level of prescription, with mandatory substantive policies being the most prescriptive and procedural, voluntary the least prescriptive.

This comparison revealed considerable variation among countries, as well as across land ownership types. The most prescriptive policies were found in public lands in developed countries while the least prescriptive addressed private land in developed countries. On average, however, developing countries had the highest levels of prescription as well as the most demanding environmental threshold requirements (McDermott *et al.*, 2010). This suggests that 1) countries further in their forest transition do not necessarily have more stringent environmental policies than less-developed countries, and 2) precisely what constitutes legal forest practice also varies considerably worldwide.

Meanwhile, whatever the content of a country's forest regulations, political and economic drivers outside the forest sector have played a major role in forest outcomes. Currently the leading driver of forest loss in Southeast Asia is the expansion of palm oil plantations, while cattle and soy farming are responsible for the majority of deforestation in Latin America. Demand for these products comes from urban markets worldwide, including growing domestic markets in Latin American and Asia as well as North America and Europe (DeFries *et al.*, 2010).

Thus as developed countries have passed their days of "peak deforestation" and are now experiencing forest regrowth, rates of deforestation in the developing world have accelerated. The last few decades have also seen the rapid rise of an international environmental movement pushing for forest conservation across the globe. The combined pressure for conservation and development has generated considerable conflict among a wide range of stakeholders worldwide.

Footprints

The solutions to this conflict depend, in part, on the precise understanding of "forest transition". If one looks only at patterns of forest change within the borders of individual countries, then one solution might be to incentivize or compensate developing countries for increasing the effectiveness of their environmental regulation and hence accelerating their forest transition. However, there is a growing body of research that provides a contrasting explanation, which we summarize here as global environmental "footprints". This research examines countries' overall resource consumption, including internationally traded goods, and finds that overall consumption continues to increase with rising GDP (Wiedmann *et al.*, 2013). While increases in domestic production efficiency may partially support this increased consumption, in general the higher the GDP, the more countries rely on imported goods, thereby displacing their environmental impacts outside of their own borders (ibid.). To the extent the "forest transition" is due to the displacement of forest loss elsewhere, this calls into question what will happen as more countries develop, and there is dwindling room for such displacement.

A footprints perspective also points to a correlation between global inequality and resource degradation. For example, Rice (2007) has found an inverse relationship between a country's per capita wood consumption and rates of forest loss. At the extreme ends, developed countries consume 18 times the forest products consumed by the least-developed countries. The theory of "ecologically unequal exchange" explains such phenomena by arguing that less-developed countries suffer from a disadvantaged position in global trade that leaves them perpetually dependent on primary resource extraction. Unable to capture an adequate share of global profits, they thus remain "stuck at the bottom" in terms of the benefits they accrue from global trade. This prevents the development of a middle class and an active civil society to push for protection of resources, and leaves governments unable to invest in environmental protection. From this perspective, in order to assess the effectiveness of policies one must consider their effect on the equality of resource distribution. Furthermore, it highlights how if all countries were to attain the levels of economic development currently enjoyed by the global North, it will not be possible to address global forest loss without also addressing rising resource consumption.

The following section will now examine the emergence of inter-governmental processes that have attempted to navigate these different perspectives on the drivers of forest change and find common ground on the governance of the world's forests. As will be clear from the following analysis, this has been a tall order indeed, but one which has grown very substantially in momentum and scope.

Securing the world's forests: The role of governments

The United Nations forest processes

Inter-governmental collaboration on forests first began in earnest in the mid twentieth century. This was a time of rapid growth in international trade (both within and outside of the forest sector) and accelerating loss of tropical forests. Concern about the effects of globalization on both forests and the forest industry in turn drove interest in international cooperation on forest governance and trade (Humphreys, 2006). This period saw the launch, for example, of the forestry department within the Food and Agricultural Association of the United Nations (FAO), an organization which has played a pivotal role in monitoring and reporting global and regional trends across the world's forests. Concern about the loss and degradation of tropical forests also spurred the 1985 Tropical Forest Action Plan (TFAP). TFAP involved a collaborative effort among national governments and development agencies to provide coordinated development assistance for sustainable tropical forest management (FAO, 1985).

Despite these early efforts, by the time of the 1992 Earth Summit in Rio, tropical forest loss had accelerated to a rate of over 15 million hectares per year (FAO, 1990). Meanwhile, a large and rapidly expanding network of international environmental NGOs were increasing pressure on governments to take firmer action. Many of these NGOs, furthermore, were concerned about unsustainable forest practices in temperate and boreal regions as well as in the tropics. This led to a proposal at the Rio Earth Summit for a global forest convention that would set international standards for sustainable forest governance worldwide (Humphreys, 2006).

Support for a global forest convention proved far from universal, however, and countries were unable to reach agreement in Rio. Instead, they established a new platform within the UN system, known as the Intergovernmental Panel of Forests (IPF) to continue the negotiations. After several years, the IPF was reformulated as an Intergovernmental Forum on Forests (IFF) and later it was shifted yet again to be a more permanent body, the UN Forum on Forests (UNFF). To date, no convention has been agreed. Instead, the UNFF has produced a "non-legally binding instrument on all types of forests" (UNFF, 2007). This instrument articulates a number of globally shared goals, including the goal to arrest global forest loss, but does not impose any obligations on country parties to meet them.

While the quest for a legally binding forest agreement has faltered, the last 20 years of negotiations have shed considerable light on what factors were determining the support or opposition of different countries to such an agreement. Humphreys (2006) has outlined several points of conflict. Early in the negotiations, developing countries have defended their sovereign right to develop their forest frontiers as they choose. Over time, the North/South divide shifted to represent a divide between those countries with major forest industries who were major exporters of timber and those who weren't. Countries with large industrial exports tended to support a global convention as a means to establish the legitimacy of their forest practices and thus protect themselves from external embargoes. International environmental NGOs who were initially supportive of a convention shifted their position, expressing concern that export-oriented countries would set too low a bar for performance and hence would legitimate business as usual. The result was a stalemate.

Regional collaboration on illegal logging

Widespread agreement over the definition of "sustainable forest management" remains elusive to this day. However, in the early 2000s a new international coalition began to form that largely bypassed the issue by reframing the problem of sustainability as one of legality. It was observed that many developing countries already had developed quite stringent forest regulations but that these regulations were not enforced. The harnessing of global energies to help states enforce their laws appeared to offer a "win–win" solution that would overcome many of the political bottlenecks that plagued negotiations over a forest convention. Firstly, it would help reinforce rather than undermine state sovereignty, and help governments capture millions of dollars in lost state revenue (World Bank, 2004). Secondly, it would provide a tool for environmental groups to fight some of the worst forest practices. Thirdly, it would help create a more "level playing field" for those in the forest industry already abiding by existing government rules and regulations (McDermott, 2014).

Speaking to its widespread appeal, the illegal logging frame has since spurred a wide range of activities among governments worldwide. This includes a series of regional Forest Law Enforcement and Governance (FLEG) processes launched with support from the World Bank. The goal of these processes is to "mobilize international commitment from both producer, consumer and governments" through ministerial partnerships in Africa, East Asia, Europe and North Asia (World Bank, 2013). It also includes an EU-level effort launched in 2003, known as the Forest Law Enforcement Governance and Trade (FLEGT) Action Plan. The FLEGT Action Plan, as signified by the "T", has focused on ways to leverage the EU's important role in the global timber trade to incentivize developing countries to enforce their laws. A core mechanism within this plan, is the establishment of "Voluntary Partnership Agreements" (VPAs) between the EU and partner countries. Part of the VPA includes a system of "legality licensing" for timber imported into the EU from its VPA partner countries. Once these licensing schemes are in place, the EU could help partner countries enforce their laws by barring the import of unlicensed timber (EC, 2003).

Another core set of illegal logging strategies are a series of unilateral laws set by key importing countries including the US, the UK and Australia. The US was the first to move on this through the expansion of its pre-existing Lacey Act. The 1990 Lacey Act was initially established to curb illegal trade in wildlife and wildlife parts. Specifically, the Act declared it illegal to import or trade domestically in wildlife or parts that violate the laws of the country of origin. In 2008, the Lacey Act was extended to include most wood products as well (USDA APHIS, 2010). Two years later, the EU passed the EU Timber Regulation (EU TR). This EU TR requires that those buyers who import wood products listed in the EU TR into the EU must demonstrate due diligence to ensure that their products do not originate from illegal practices (EC, 2010). The 2012 Australian Illegal Logging Prohibition Act follows a similar logic for imports into Australia (Australia, 2012).

Procurement policies are another way that governments have engaged with the issue of legality as well as sustainability in wood supply chains. These policies involve requirements or preferences for wood products consumed by government institutions. Since governments are very large consumers, their procurement policies can have significant impact across the supply chain. In the US, for example, federal, state and local government purchasing amounts to nearly 20% of the country's GDP (RCA, 2002). While under the rules of the World Trade Organization (WTO), governments are not allowed to institute universal restrictions on imports based on environmental requirements for production, they are allowed to include such specifications in their role as major consumers.

Initially the focus of many government procurement policies for wood products was on ensuring that such products came from sustainable or certified sustainable sources. With the rise of the illegal logging initiatives, these requirements were expanded to directly address the issue of illegality. For example, the UK Central Point of Expertise on Timber Procurement (CPET) developed a tiered system for screening wood purchased by the UK government. One tier addressed requirements that the wood be verified as legal (CPET, 2006b) and the other outlined preferences for wood certified as sustainable by certification schemes deemed to meet CPET requirements (CPET, 2006a). In this latter case, the UK government appears to have influenced the design of the certification schemes themselves as subsequent changes were made to these schemes that aligned with UK government requirements (Overdevest, 2010).

The net effect of these multiple illegal logging and procurement initiatives is yet to be determined. However, one indication of their importance to exporting countries is the large number of countries that are entering into negotiations over FLEGT VPAs, or considering doing so. There are currently six countries in Africa and one in Southeast Asia (Indonesia) that have completed and signed VPA agreements. Another nine, including one in Central America and one in South America, who are in the process of negotiating VPAs, as well as 11 who have formally expressed interest (EFI, 2014). As of yet, none of these countries have fully operational legality licensing schemes that are formally recognized by the EU. However, many have made progress at the domestic level. For example, Indonesia has instituted across-the-board requirements that all wood producers must undergo third party audits by licensed auditing organizations if they are to sell wood legally either within or outside of the country (EU *et al.*, 2011).

While there is thus substantial evidence that these initiatives are triggering changes in state governance in some countries, it is much more difficult to assess their broader effects on sustainability. The reasons for this are many. Firstly, an analysis of the wood products trade worldwide, highlights the increasing importance of China, India and other countries who are not as actively involved in international illegal logging initiatives. For example, China is the world's largest importer of tropical logs (McDermott and Cashore, 2009). The existence of markets in China and elsewhere that do not require legality verification, could result in "leakage" whereby illegal logs are simply diverted to these less-discriminating markets.

On the other hand, "spill-over" effects are also possible, due to the interconnected nature of global trade. For example, much of the tropical wood imported into China is used to create furniture and other secondary processed wood products that are then re-exported into the US and EU (McDermott and Cashore, 2009). This can place pressure on China to address the Lacey Act, EU TR and other such initiatives. Furthermore, as noted by Vogel

(1997) and others, industrial producers with operations in multiple countries face incentives to standardize their approach to legality and may themselves pressure governments in other jurisdictions to support these efforts as a means to enhance their competitive advantage. This phenomenon has led some to hypothesize that increasing globalization of the wood trade may foster a "ratcheting up", rather than a "race to the bottom" of environmental regulations (Cashore *et al.*, 2007).

Regardless of the degree of leakage and/or spill-over, the core issue for sustainability lies in their on-the-ground impacts. Assessing these impacts requires measuring the effects of illegal logging initiatives on forests and people, including whether they are addressing the true drivers of forest change, how they are impacting environmental and social inequalities, and whether they are effective in tackling unsustainable global consumption. Here the evidence is much scantier and not as encouraging. If we adopt a global footprints perspective that inequalities in global trade are hampering some countries' abilities to develop effective domestic forest governance, then it is important to consider how these legality licenses will affect competitive advantages in global markets. Since illegal logging is deemed a greater problem in developing countries then it is reasonable to expect that developing countries will have more difficulty addressing legality requirements and/or convincing buyers that they have done so. In other words, legality initiatives are likely to reinforce global trade inequalities, at least in the short to medium term.

Legality initiatives are also affecting equity issues at the domestic level. On the one hand, research has found that FLEGT VPA processes have contributed to a more democratic process of forest policy-making, at least during their negotiation phase. On the other hand, they have negatively impacted small and medium enterprises that lack the land tenure security and economies of scale to meet requirements for third party auditing (Lesniewska and McDermott, 2014).

The impacts of timber legality initiatives are also limited by their focus on a single sector. By focusing only on wood production, they do not address the primary drivers of forest loss, i.e. the expansion of commercial agriculture. Constraining access to markets for timber, without addressing these other drivers, may reduce the profitability of maintaining land in forest and thus spur deforestation.

REDD+

Meanwhile, in 2007, a new initiative emerged that promised to directly address the cross-sectoral nature of forest change. This breakthrough was achieved through the linkage of forests with climate via Reducing Emissions from Deforestation and Degradation and forest enhancement[1] (REDD+), a mechanism under the United Nations Framework Convention on Climate Change (UNFCCC). At this time, it was estimated that tropical forest loss was accounting for some 18% of total global greenhouse gas emissions

(Stern, 2007). Furthermore, based on calculations of the "opportunity cost" of foregoing agricultural expansion in tropical forest frontiers, it appeared that reducing emissions from forest loss offered a particularly fast and cheap way to reduce global emissions (Eliasch, 2008). The core logic of REDD+ was that it should be voluntary, and that developed countries would help pay those developing countries that chose to participate for the opportunity costs of reducing deforestation within their borders. This would be achieved through "performance based payments" for measured reductions in the emission of forest carbon resulting from forest loss and degradation (UNFCCC, 2011). The term "performance based" signifies that once REDD+ is fully operational, payments will be based strictly on the size of a country's verified emissions reductions not the cost of its efforts to achieve such reductions.

Thus unlike other international forest initiatives before it, REDD+ was not focused on timber production or timber trade, but promised to address all of the core forest drivers both within and outside of the sector. It also promised to overcome a range of other stumbling blocks that have stymied global agreements in the past. Firstly, it respected countries' sovereign rights to develop their forest frontier by making participation voluntary. It also promised compensation for forest conservation activities that impeded this development. Government sovereignty was even further enhanced by the decision that REDD+ accounting under the UNFCCC would be focused at the national scale, and fall under the authority of national governments (UNFCCC, 2011).

The implementation of REDD+ was intended to occur in three phases. Firstly, countries were to develop national REDD+ strategies or action plans, put in place enabling policies and legislation, and build capacity. Implementation would then start under the second phase, which involves results-based demonstration activities along with further training, capacity-building and technology development and transfer. After the first two REDD+ phases, collectively referred to as "REDD+ readiness" activities, countries will have prepared their carbon "reference levels" or "reference emissions levels" that set the baseline for monitoring forest carbon from that point onwards. The calculation of baselines promises to be controversial, since countries face incentives to exaggerate their estimation of "business as usual" carbon emissions in order to maximize the payments they receive for reducing it. Nevertheless, agreement on a baseline is essential to enable the third and final phase of REDD+. From that point onwards, REDD+ is to be "performance-based" and tied to a robust system of measurement, reporting and verification (MRV) of forest carbon (UNFCCC, 2011). No timeline has been established for countries to pass through the three phases of REDD+. Rather, it is anticipated that countries will differ in the time needed to prepare and qualify for performance-based payments (e.g. Eliasch, 2008).

Many international NGOs and various other stakeholders have expressed concerns that the exclusive focus of REDD+ accounting on carbon will negatively impact other forest values. Tropical forests are uniquely rich in biodiversity and support diverse populations of indigenous and local communities. This has led to fears that transforming forest carbon into a globally tradable commodity might lead to "land grabs" that displaced local people from their traditional territories and livelihoods. It also could lead to the conversion of natural forests to plantation-based "carbon farms" with profoundly negative effects on biodiversity. These concerns led to the inclusion of "safeguards" in the 2010 UNFCCC REDD+ agreement. The safeguards state that REDD+ activities should respect local rights and knowledge and support local livelihoods, and also should not lead to the conversion of natural forests to plantations but should enhance biodiversity (UNFCCC, 2011, Appendix 1).

As of the 2013 UNFCCC Conference of Parties in Warsaw, it has been decided that countries must report on how they are respecting the REDD+ safeguards as a pre-requisite to receiving REDD+ finance. However, the question of precisely how this finance will be generated and managed is still not fully resolved. The Green Climate Fund (GCF) is to play a coordinating function. It has also been decided that REDD+ funding may come from "a variety of sources, public and private, bilateral and multilateral" (UNFCCC, 2014).

To date, the majority of international funds for REDD+ readiness have been channeled through the World Bank and various regional development banks, and/or provided by a range of UN agencies and private actors. These various funding agencies, furthermore, have helped shape expectations for how countries should prepare for REDD+. For example, funding from the World Bank's Forest Carbon Partnership Facility (FCPF) hinges progressively on a readiness preparation proposal (R-PP), followed by an emission reduction program idea note (R-PIN) and the establishment of a monitoring and evaluation system (World Bank and UNDP, 2014). All of these planning stages must, among other things, ensure compliance with the World Bank's internal environmental and social safeguards. Countries that have received funding from multiple sources may face multiple layers of such requirements.

In addition, other sets of rules around REDD+ have emerged within the context of private and largely unregulated markets for forest carbon (Peters-Stanley and González, 2014). These include carbon certification schemes, such as the Verified Carbon Standard (VCS) and the Climate, Community and Biodiversity Alliance (CCBA). The CCBA has also teamed up with CARE international to partner with REDD+ governments to assist them in developing national "safeguard information systems" in compliance with UNFCCC requirements. This initiative, known as the REDD+ Social and Environmental Safeguards (REDD+ SES) represents a form of public/private "hybrid" governance shaping how REDD+ is operationalized (McDermott *et al.*, 2012). It also speaks to the overall complexity of actors and initiatives, and diverse sources of authority, that are becoming increasingly typical of international forest governance.

While REDD+ has clearly generated a wide range of activity at the international level, its implementation has proven much less "quick and easy" than some originally envisioned. Among the many challenges that REDD+ is currently facing, is that the finance required to operationalize REDD+ at a scale that would influence global climate has yet to materialize. Furthermore, it is likely that raising and maintaining the necessary finance will be contingent on the new global climate agreement anticipated to take effect in 2020. The setting of ambitious emissions reduction targets by participating countries may be required to generate the level of finance needed to move REDD+ to scale.

Meanwhile, the promise of REDD+ has spurred a wide range of actions at multiple scales that are arguably affecting forest governance, if in different ways than initially envisioned. For example, there is already evidence that the REDD+ safeguards have influenced the legal frameworks governing forests in REDD+ countries. In Mexico, a revision to an environmental and forest law legally enshrines requirements equivalent to many of the REDD+ safeguards (Mexico, 2012). In Indonesia, REDD+ appears to have catalyzed a recent court case ruling that indigenous rights remain unextinguished across vast areas of ungazetted forestland (Ituarte-Lima *et al.*, 2014).

To date, a large proportion of the funding for REDD+ readiness has been invested in technologies for measuring, reporting and verifying forest change. This has likewise affected governments' forest strategies. For example, Brazil, the country that until recently was responsible for the largest annual forest loss of any country worldwide, quartered its deforestation rate between 2006 and 2013 (Nepstad *et al.*, 2014). This has been achieved through a variety of initiatives, including the use of satellite monitoring coupled with increased law enforcement to halt illegal forest clearance, industry-supported moratoriums on the expansion of soy, and systems for tracking timber legality (McDermott *et al.*, 2014). While these activities are not explicitly linked to REDD+, it is not unlikely that such actions are partly incentivized by the large levels of global attention and resources that have been focused on REDD+. Meanwhile Indonesia, another historical world leader in tropical deforestation, is undertaking a major effort known as "One Map". One Map is an attempt to regularize land tenure across sectors, resolving extensive problems of overlapping boundaries among forest, agriculture and mining concessions. While One Map is not a REDD+ initiative per se, it is consistent with the emphasis of REDD+ on more accurate monitoring, accounting and monetizing of forest carbon. It has also been enabled by significant shifts in the balance of forestry decision-making power that occurred during the development of Indonesia's national REDD+ strategy (Mulyani and Jepson, 2013).

Cross-sectoral strategies

Yet, while some developing countries are making headway in slowing deforestation, forest loss has accelerated elsewhere, and new frontiers for palm oil and soy are opening up to feed rising global demand. Currently world prices for forest carbon are far below the average opportunity costs of clearing forests for the agricultural crops responsible for the majority of deforestation

(Butler *et al.*, 2009; Börner and Wunder, 2008; Peters-Stanley and González, 2014). Meanwhile government policies supporting the use of biofuels may further incentivize oil crop expansion (Overmars *et al.*, 2011). There have been a number of recent multilateral as well as unilateral initiatives attempting to address the impacts of these markets on forests. For example, the EU has passed environmental standards for biofuel consumption (EC, 2009). The EU has also commissioned a study to assess the quantity of deforestation "embodied" in products imported into the EU (Cuypers *et al.*, 2013). Meanwhile unilateral actions have also been taken. For example, the UK has recently issued a statement pledging to buy only "credibly certified sustainable palm oil" by the end of 2015 (UK, 2014).

This increasing focus outside of the forest sector on other commercial supply chains could be seen as a "new wave" of landscape-level forest-related policy. However, to date, the core focuses of most of these policies are sector specific, i.e. focused on forest carbon, or biofuels, or palm oil. The risk remains that targeting each sector separately may displace impacts both geographically and across sectors while doing little to address overconsumption. Meanwhile, global forest frontiers remain highly dynamic and diverse threats continue to emerge, such as recent and major expansions of hydro-electric projects and the proliferation of both large- and small-scale mining (Fearnside *et al.*, 2012; Sonter *et al.*, 2014). Table 3.1 summarizes the above analysis of core governmental initiatives aimed at international forest governance.

Conclusions and policy recommendations

The above analysis maps out the increasingly complex landscape of state-based forest and forest-related governance to date. The analysis began by considering the importance and changeability of how governments and other stakeholders have framed the core challenges that forest policies aim to address. Since the advent of industrialization, governments in the early stages of industrial development have frequently emphasized forest exploitation and the conversion of forests to other uses. However, over the last century, deforestation in developed countries has reversed, and forest protection has gained prominence in international discourse. This has created tensions among developed and developing countries over the right of individual nations to benefit from the economic development of their forest frontiers. It has also raised questions about our underlying assumptions regarding who and what is driving global forest change, and what strategies governments should adopt to address it.

The literature on "forest transitions" has focused on changes in forest cover within the borders of individual countries. Developed countries emerge as environmental leaders from this perspective, having outgrown their earlier periods of forest loss and exploitation and even achieving recent gains in forest cover. From this perspective, the challenge is to bring developing countries up to the same level of performance. Arguably most of the initiatives discussed above are in line with this perspective. The focus on

Table 3.1 Summary of core governmental initiatives aimed at international forest governance[a]

Target theme	*Focus area for achieving a target theme*		
	Forests	*Wood products*	*Forest change drivers*
Sustainable forest management	UN Forum on Forests (UNFF), Non-Legally Binding Instrument on all types of forests (NLBI)	Government procurement policies	
Combatting illegal logging		Regional Forest Law Enforcement and Governance (FLEG) processes; EU Forest Law Enforcement, Governance and Trade (FLEGT) Action Plan; 2008 US Lacey Act Amendment; 2010 EU Timber Regulation; 2012 Australian Illegal Logging Prohibition Act; government procurement policies	
Mitigating carbon emissions	Reducing Emissions from Deforestation and Degradation (REDD+)		Reducing Emissions from Deforestation and Degradation (REDD+)
Abating conversion of forests to agricultural land			EU study on embodied deforestation; EU sustainable biofuels policy; UK declaration on palm oil

[a] This table lists governmental initiatives deemed to be particularly influential at a global scale. For a more comprehensive list of international forest initiatives see, for example: McDermott *et al.* (2007)

excluding illegal logging from world trade has been justified based in part on the argument that it will enhance the environmental and social performance of forest production in developing countries. Likewise, the REDD+ mechanism is focused on paying developing countries to forgo development of the forest frontier in favor of forest conservation.

A perspective that focuses on global "footprints", however, frames the problem somewhat differently. This perspective highlights how per capita material consumption continues to increase along with economic growth, and is currently exceeding "planetary boundaries". It highlights how developed countries have succeeded in protecting their forests in part by displacing deforestation elsewhere, and how this has led to high levels of global inequity. It illustrates how the core drivers of forest loss lie in agriculture, not the forest sector, and are the result of rising urban consumption in both developed and developing countries.

The forest governance approaches outlined in this chapter, and summarized in Table 3.1, appear relatively less suited to address this second version of the problem. The initiatives focused on the regulation of the timber trade fail to address the role of agriculture in driving forest loss, and in fact risk encouraging forest conversion by making wood production even less profitable than agricultural production. They also may increase global inequalities of access to timber markets and inhibit the distribution of its benefits among local communities. REDD+, on the other hand, does attempt to address the agricultural drivers of forest change. However, it relies on high prices for forest carbon. In part because these high prices have failed to materialize, REDD+ has yet to be implemented at the scale needed to arrest forest loss and significantly mitigate global emissions. Nevertheless, the promise of billions of dollars being channeled through "results-based" payments has directed much of the financial investment into the technical challenge of measuring and monitoring forest carbon. Meanwhile, none of the above instruments tackle the issue of unsustainable rates of consumption.

Despite all of the shortcomings of governmental and inter-governmental efforts to date, it would be a mistake to dismiss them as failures. Rather these initiatives remain highly dynamic and continually evolving, and this has produced many positive unintended, indirect or spill-over effects. For example, both the VPA and REDD+ processes appear to have opened political space for new actors at international and local levels to participate in forestry decision-making. Likewise, REDD+ has helped incentivize countries to develop parallel, nationally driven processes to arrest forest change (e.g. in Brazil) and/or safeguard indigenous rights (e.g. in Mexico and Indonesia).

So what, then, is the best way forward? Firstly, we can learn that it is important to adopt a holistic understanding of "sustainability" in the context of forests. This perspective would consider the cross-sectoral nature of forest change (i.e. the role of agricultural crops such as palm oil and soy in driving forest loss) as well as the role of global inequalities in trade and consumption (i.e. how unsustainable rates of consumption in developed countries contribute to unsustainable resource extraction in poor countries). Secondly, we can learn of the importance of a holistic understanding of policy landscapes, that takes into account the interactions of policies at multiple scales, both inside and outside of the forest sector (i.e. how negotiations for a new international climate agreement are influencing domestic forest policies). Thirdly, and in a related way, we can learn of the importance of holistic and long-term evaluation. That is, rather than rely solely on a result-based assessment of short-term impacts of any one intervention in any single sector, it is important to look at the suite of interventions together, and to consider not only intended outcomes but also unintended, indirect and longer-term effects.

This way forward would require space for a wide diversity of approaches. It would also require the design of strategies specifically tailored to address the current shortcomings of international forest governance. Among these are the ways in which international demands for legality verification (e.g. FLEGT) or carbon monitoring (e.g. REDD+) lead to changes in domestic forest governance that disadvantage small-scale producers or farmers without secure land tenure. This is not simply a matter of raising finance to help smallholders meet increasing legal requirements. Such approaches are unlikely to provide sustainable solutions at scale. What is needed instead, is to prioritize the development of new initiatives that are tailored specifically to support sustainable local production and trade. These initiatives would likely place less emphasis on complex planning, and prescriptive regulatory and licensing procedures designed to satisfy external market actors, but rather support local communities and entrepreneurs in designing new local and community-based governance systems.

Strategies are also needed that focus on equitable ways of reducing consumption that do not displace overconsumption of one product with another. This is the least-developed area of forest policy and it is worthy of much greater attention. Precisely how this might be done will vary by country and context. But the first step towards achieving it is better acknowledging and understanding our forest-related footprints and prioritizing equitable solutions to address it.

Acknowledgments

This chapter was produced with support from the European Union's Seventh Programme for research, technological development and demonstration under grant agreement No. 282887, Future-oriented integrated management of European forest landscapes (INTEGRAL).

Note

1 Forest activities included under REDD+ are: 1) reducing emissions from deforestation; 2) reducing emissions from forest degradation; 3) conservation of forest carbon stocks; 4) sustainable management of forests; 5) enhancement of forest carbon stocks (UNFCCC 2011: C.70). Activities 3)–5) constitute the "+" in REDD+.

References

Australia. 2012. Illegal Logging Prohibition Act, No. 166.

Barbier, E., Burgess, J. and Grainger, A. 2010. The forest transition: towards a more comprehensive theoretical framework. *Land Use Policy*, 27, 98–107.

Börner, J. and Wunder, S. 2008. Paying for avoided deforestation in the Brazilian Amazon: from cost assessment to scheme design. *International Forestry Review*, 10, 496–511.

Butler, R.A., Koh, L.P. and Ghazoul, J. 2009. REDD in the red: palm oil could undermine carbon payment schemes. *Conservation Letters*, 2, 67–73.

Cashore, B., Auld, G., Bernstein, S. and McDermott, C. 2007. Can non-state governance "ratchet up" global environmental standards? Lessons from the forest sector. *Review of European Community and International Environmental Law*, 16, 158–172.

CPET. 2006a. UK Government Timber Procurement Policy: Criteria for Evaluating Certification Schemes (Category A Evidence). Central Point of Expertise on Timber.

CPET. 2006b. UK Government Timber Procurement Policy: Framework for evaluating Category B Evidence. Central Point of Expertise on Timber.

Cuypers, D., Geerken, T., Gorisson, L., Lust, A., Peters, G., Karsten, J., Prieler, S., Fisher, G., Hizsnyik, E. and Van Velthuizen, H. 2013. *The impact of EU consumption on deforestation: Comprehensive analysis of the impact of EU consumption on deforestation*. EU.

DeFries, R., Rudel, T., Uriarte, M. and Hansen, M. 2010. Deforestation driven by urban population growth and agricultural trade in the twenty-first century. *Nature Geoscience: Letters*, 3, 178–181.

EC. 2003. Forest Law Enforcement, Governance and Trade (FLEGT) Proposal for an EU Action Plan. European Commission.

EC. 2009. Directive 98/70/EC as regards specification of petrol, diesel and gas-oil and introducing a mechanism to monitor and reduce greenhouse gas emissions and amending Council Directive 1999/32/EC as regards the specification of fuel used by inland waterway vessels and repealing Directive 93/12/EEC. European Commission.

EC. 2010. Regulation (EU) No. 995/2010 of the European Parliament and of the Council of 20 October 2010 laying down the obligations of operators who place timber and timber products on the market. Text with EEA relevance. European Commission.

EFI. 2014. EU FLEGT Facility > Voluntary Partnership Agreements [Online]. European Forest Institute. Available: www.euflegt.efi.int/vpa-countries. Accessed June 18 2014.

Eliasch, J. 2008. *Eliasch Review: Climate change: financing global forests*. UK Office of Climate Change.

EU, Indonesia and DFPPM. 2011. FLEGT Voluntary Partnership Agreement between Indonesia and the European Union Briefing Note.

FAO. 1985. Tropical Forestry Action Plan. Rome, Italy: Committee on Forest Development in the Tropics.

FAO. 1990. Forest Resources Assessment 1990 – Global Synthesis. Rome.

FAO. 2013. State of the World's Forests 2012.

Fearnside, P.M., Figueiredo, A.M.R. and Bonjour, S.C.M. 2012. Amazonian forest loss and the long reach of China's influence. *Environment, Development and Sustainability*, 15, 257–263.

Humphreys, D. 2006. *Logjam: Deforestation and the Crisis of Global Governance*. London and Sterling, VA, Earthscan.

Ituarte-Lima, C., McDermott, C.L. and Mulyani, M. 2014. Assessing equity in national legal frameworks for REDD+: the case of Indonesia. *Environmental Science and Policy*, 44, 291–300.

Lesniewska, F. and McDermott, C.L. 2014. FLEGT VPAs: laying a pathway to sustainability via legality: lessons from Ghana and Indonesia. *Forest Policy and Economics*, doi 10.1016/j.forpol.2014.01.005.

Mather, A.S. 1992. The forest transition . *Area*, 24, 367–379.

McDermott, C.L. 2014. REDDuced: From sustainability to legality to units of carbon – The search for common interests in international forest governance. *Environmental Science and Policy*, 35, 12–19.

McDermott, C. and Cashore, B. 2009. *Forestry Driver Mapping Project: Global and US Trade Report*. New Haven: Global Institute of Sustainable Forestry, Yale University.

McDermott, C., O'Carroll, A. and Wood, P. 2007. *International Forest Policy – the instruments, agreements and processes that shape it*. United Nations Forum on Forests, UN Department of Economic and Social Affairs.

McDermott, C.L., Cashore, B. and Kanowski, P. 2010. *Global Environmental Forest Policies: An International Comparison*. London: Earthscan.

McDermott, C.L., Coad, L., Helfgott, A. and Schroeder, H. 2012. Operationalizing social safeguards in REDD+: actors, interests and ideas. *Environmental Science and Policy*, 21, 63–72.

McDermott, C.L., Irland, L.C. and Pacheco, P. 2014. Forest certification and legality initiatives in the Brazilian Amazon: lessons for effective and equitable forest governance. *Forest Policy and Economics*, 20, 1–9.

Mexico 2012. Reforms to the Ley General del Equilibrio Ecológico y la Protección Ambiente y de la Ley General de Desarrollo Forestal Sustentable.

Mulyani, M. and Jepson, P. 2013. REDD+ and forest governance in Indonesia: a multistakeholder study of perceived challenges and opportunities. *The Journal of Environment and Development*, 22, 261–283.

Nepstad, D., McGrath, D., Stickler, C., Alencar, A., Azevedo, A., Swette, B., Bezerra, T., Digiano, M., Shimada, J., SeroaDa Motta, R., Armijo, E., Castello, L., Brando, P., Hansen, M.C., Mcgrath-Horn, M., Carvalho, O. and Hess, L. 2014. Slowing Amazon deforestation through public policy and interventions in beef and soy supply chains. *Science*, 344, 1118–1123.

Overdevest, C. 2010. Comparing forest certification schemes: the case of ratcheting standards in the forest sector. *Socio-economic Review*, 8, 47–76.

Overmars, K.P., Stehfest, E., Ros, J.P.M. and Prins, A.G. 2011. Indirect land use change emissions related to EU biofuel consumption: an analysis based on historical data. *Environmental Science and Policy*, 14, 248–257.

Peters-Stanley, M. and González, G. 2014. *Sharing the Stage: State of the voluntary carbon markets 2014: Executive Summary*.

RCA. 2002. *Using Less Wood: Focus on Government Purchasing*. Washington DC: Resource Conservation Alliance.

Rice, J. 2007. Ecological unequal exchange: consumption, equity, and unsustainable structural relationships within the global economy. *International Journal of Comparative Sociology*, 48, 43–72.

Rudel, T., Defries, R., Asner, G.P. and Laurance, W. 2009. Changing drivers of deforestation and new opportunities for conservation. *Conservation Biology*, 23, 1396–1405.

Sonter, L.J., Moran, C.J., Barrett, D.J. and Soares-Filho, B.S. 2014. Processes of land use change in mining regions. *Journal of Cleaner Production*, 84, 494–501.

Stern, N. 2007. *Stern review on the economics of climate change: Executive summary*. UK Department of Energy and Climate Change.

UK. 2014. UK statement on palm oil. The Central Point of Expertise for Timber Procurement (CPET).

UNFCCC. 2011. The Cancun Agreements Dec. 1/CP.16. United Nations Framework Convention on Climate Change.

UNFCCC. 2014. Report of the Conference of the Parties on its nineteenth session, held in Warsaw from 11 to 23 November 2013: Part two. United Nations Framework Convention on Climate Change.

UNFF. 2007. Non-legally binding instrument on all types of forests. United Nations Forum on Forests.

USDA APHIS. 2010. Lacey Act Primer. Washington DC: United States Department of Agriculture, Animal and Plant Health Inspection Service.

Vogel, D. 1997. Trading up and governing across: transnational governance and environmental protection. *Journal of European Public Policy*, 4, 556–571.

Wiedmann, T.O., Schandl, H., Lenzen, M., Moran, D., Suh, S., West, J. and Kanemoto, K. 2013. The material footprint of nations. *Proceedings of the National Academy of Sciences*, 112, 6271–6276.

World Bank. 2004. *Sustaining Forests: A Development Strategy*. Washington, DC: World Bank.

World Bank. 2013. Regional Forest Law Enforcement and Governance (FLEG) Initiatives [Online]. Available: ww.worldbank.org/en/topic/forests/brief/fleg-regional-forest-law-enforcement-governance.

World Bank and UNDP. 2014. Forest Carbon Partnership Facility (FCPF) | CFO [Online]. Available: www.climatefinanceoptions.org/cfo/node/57.

4 Achieving sustainability through market mechanisms

Benjamin Cashore, Chris Elliott, Erica Pohnan, Michael Stone and Sébastien Jodoin

For over a generation, government, business, and non-governmental organizations have been increasingly turning to "market mechanisms" to promote sustainable forestry around the world. In contrast to traditional "command and control" approaches in which governments require and enforce a specified behavior through mandatory regulations (Taylor *et al.*, 2012) market mechanisms attempt to create economic incentives for firms, managers and individuals, to behave in ways that improve environmental stewardship and foster social values. Given that many have criticized previous efforts to combat deforestation for falling short of intended goals, the role and use of market mechanisms has become of great interest to practitioners, scholars and policy makers (McDermott, 2014a).

The direct impact of market mechanisms is not always easy to measure and is rarely immediate (Auld and Cashore, 2012). Market mechanisms also interact with public policies in myriad ways. Depending on design and trajectory, these efforts can lead to standards that "ratchet up," "ratchet down," or produce status quo forestry practices. The purpose of this chapter is to shed light on these processes by exploring the multiple pathways through which market mechanisms exert influence.

In this chapter, we define "market mechanisms" as interventions or policy instruments that attempt to harness market incentives towards addressing specified policy objectives. This focuses attention on policy instruments that draw on some type of economic "self-interest" of firms, individuals or communities as a means to promote socially and environmentally responsible stewardship (Perman, 2003). As we detail below, the broad interest in providing economic incentives has resulted in myriad distinct approaches, from imposing taxes on pollution or resource use, providing subsidies or payments, creating tradable emissions resource permits, to mandating voluntary certification standards or codes of conduct. Supporters note their capacity to encourage innovation in the achievement of policy ends and to generate support and compliance from those whose behaviors are the focus of government policy (Gunningham *et al.*, 1998). Critics argue that they can produce suboptimal results, such as favoring powerful interests over marginalized peoples and biodiversity conservation.

To better understand their potential and limitations, we take the view that practitioners and scholars must understand the multiple logics through which a particular market mechanism might influence outcomes. To further this understanding, we develop a two-pronged analytical framework that helps us to distinguish four key dimensions with which to characterize market mechanisms, and the causal influence pathways on which they might be embarked.

Market mechanisms: Four dimensions

We identify four key dimensions that help characterize, and distinguish, what are generally referred to as "market mechanisms": (1) Private vs. public authority: from what source does the mechanism draw authority? (2) Compliance efficiency: are transaction costs reduced? (3) Economic payments: are economic payments made in exchange for goods or services provided? (4) Multi-level governance: does the mechanism need to interact with other levels of government in order to achieve impact?

Public vs. private authority

Market mechanisms can be drawn upon to build new forms of "private governance" that sit alongside or even replace traditional government-led approaches. For example, private eco-labeling systems gain their authority from consumers along global supply chains who demand products that conform to higher social or environmental standards. In these cases, market mechanisms create new forms of "non-state market driven" (NSMD) governance that unite communities along global supply chains. While these efforts draw authority from market transactions, their approach to standards development often mimics traditional regulatory approaches. For example, most eco-labeling programs identify and clearly prescribe a range of behaviors to which producers must comply in order to be "certified" as eco-friendly. These standards often appear similar to "command and control" regulatory approaches. The difference is that, unlike government regulations, private authority can only be effective to the extent that its benefits (price premium, market access, or social licenses to operate) for producers are higher than the costs of compliance with the standards.

Compliance efficiency

Governments also use market mechanisms to reduce the costs of compliance with regulations. For example the US government reduced pollution from acid rain in the 1980s by allowing individual firms to "trade" their pollution quotas with other firms, rather than regulating the specific levels of pollution from individual firms (Joskow and Schmalensee, 1998). This approach mandates an overall required outcome (not exceeding a specified level of pollution) but the market mechanisms let the market decide the most efficient response (firms burdened with costly changes to meet their quota could instead trade with firms able to reduce pollution at lower costs). This approach inspired the global and domestic "cap and trade" approach to reducing carbon emissions espoused by many economists (Betsill and Hoffmann, 2009). However, application of these instruments has led some economists to argue that answers to the *efficiency* question also inform *whether* a problem can be addressed. Scholars and practitioners must be aware that when assessing economic instruments, value judgments about problems can sometimes be presented as rational scientific approaches to problem-solving (Levin *et al.*, 2012).

Economic payments for specified behavior

"Payments for ecosystem services" (PES) in which resource users are paid to stop, or engage in, a specific behavior (Grieg-Gran *et al.*, 2005) are another type of market mechanism. For example, state forest agencies often offer cash payments to farm owners to remove portions of their land from agricultural development (Lewis, 2001). The effects of such policies will depend on whether payments for conservation are higher than the opportunity cost of planting crops. In other cases, governments mandate an actual outcome (such as a specific amount of land placed under conservation) but instead of enforcing the law through fines or incarceration, will financially reward those who follow legal requirements. Since achievement of desired outcomes often depends on the amount of resources available, it does not generally matter whether the resources come from private authority or from traditional governments: what matters more is the type, amount, and durability of the economic payment.

Interaction across multiple levels of governance

A key distinction is whether the market mechanism has the potential to influence policy change directly, or whether it must interact with other levels and/or forms of governance to achieve "on-the-ground" results (McGinnis, 1999,). For example, many developed-country government funds supporting REDD+ depend on the support of a government in a developing country to initiate policies aimed at reducing deforestation. Likewise, "legality verification" efforts that seek to identify and label legal products along global supply chains are designed to champion and reinforce existing government policies. This is different from eco-labeling systems, which have the potential to achieve results directly without government intervention. When interactions are required, analysts and practitioners must understand how market mechanisms interact with traditional public policies to create and shape behavioral change. While "direct" market-influence mechanisms such as forest certification also interact with public policies, they differ from interaction-dependent mechanisms because their potential influence is not predicated only on these interactions.

Four pathways for policy influence

In addition to the need to better distinguish among these four dimensions of market mechanisms, we argue that it is also useful to assess how and to what extent market mechanisms not only draw upon economic incentives but can also operate through other pathways of influence. Attention to these multiple pathways requires differentiating economic "globalization" from internationalization (Bernstein and Cashore, 2000). Whereas economic globalization denotes economic interdependence that is often asserted to have a "downward" influence on environmental and social standards as countries compete for capital, attention to these pathways of "internationalization" allows us to explore the role of global market instruments in fostering "upward" domestic policies and practices. Bernstein and Cashore (2000) highlighted four pathways with different influence logics. In what follows, we provide a brief overview of these pathways and offer examples from the field of global forest governance.

Markets

The *markets pathway* focuses on causal mechanisms that create behavioral and policy changes owing to some type of market incentive or disincentive. For instance, forest certification schemes that operate as non-state market driven systems (NSMD) provide additional incentives for policy makers and companies to support regulatory standards as a result of market pressures (Cashore *et al.*, 2004). There have also been a number of attempts aimed at restricting access to lucrative markets as a way to encourage the sustainable management of forests. Restrictions can come in many forms, from procurement policies that favor third-party certification, forest "legality verification" requirements enacted by consumer countries, and the use of boycott campaigns directed by NGOs. Application of the markets pathway approach directs scholars and practitioners to assess the degree of pressure, and potential "stickiness." For example, research has found that while boycotts are often useful for "agenda setting," (Sasser *et al.*, 2006) they are often short lived unless matched by more durable efforts within the markets pathway (Elliott, 2005).

International rules

The *international rules pathway* focuses on the role of international agreements, including treaties and prescriptions, in shaping domestic forest policy responses. The causal logic that explains the influence of international rules lies in their capacity to be understood as binding by the actors to whom they are addressed. This requires attention not only to international global forest policies, but also to multilateral environmental agreements (MEAs) such as the Convention on Biological Diversity (CBD) or the Convention on International Trade in Endangered Species of Wild Fauna and Flora (CITES). This pathway's relevance for market mechanisms is two-fold. First, international rules often are developed to facilitate economic global integration, such as the World Trade Organization (WTO). Hence, those promoting market mechanisms often shape what types of international agreements will be given the most attention while limiting more prescriptive "command and control" efforts such as a global forest convention. Second, the international rules pathway can be drawn on to facilitate market mechanisms, recognition of which justified our attention to the European Union's Forest Law Enforcement and Governance and Trade (FLEGT) efforts to recognize legal trade in forest products, as well facilitating agreements to enhance payments under REDD+ in developing countries.

We also note that the standards set by market mechanisms may influence the development and elaboration of international rules. For instance, experimentation with the implementation of market mechanisms in voluntary forest carbon projects in developing countries played a key role in shaping the negotiations for the adoption of rules relating to REDD+ under the UNFCCC in the second half of the 2000s (Jodoin, 2015).

Norms

The *norms pathway* serves to focus attention on assessing the role of market mechanisms in shaping global norms about sustainable forest policies. At a general level, Bernstein has found that a focus on market mechanisms has been dominant ever since the 1992 Rio Earth Summit – so much so that these "neo-liberal" norms limit, and influence, the range of policy interventions governments have at their disposal (Bernstein, 2001). At a more specific level, market mechanisms have had profound influence on the generation of problem definitions and instrument design. For example, the norm of "high conservation value forests" (HCVF) developed within the deliberations of the Forest Stewardship Council (FSC) (Hamzah *et al.*, 2007) has influenced national-level forest practices and reforms. At other times, market mechanisms may come to embrace emergent norms. Many argue that powerful norms regarding forest livelihoods, indigenous rights, and "subsidiary" governance principles, explain why public and private actors working on the operationalization of REDD+ must take into account non-carbon social benefits that recognize and protect the rights of Indigenous Peoples and local communities.

Direct access

Finally, the *direct access pathway* focuses attention on the role of market mechanisms in fostering capacity building, transferring knowledge and technology, and altering domestic power dynamics among differing interests.

Many market instruments are frequently deployed through or alongside other interventions that aim to prepare domestic actors for their implementation. For instance, several international organizations, bilateral aid agencies, and private actors have created programs of capacity-building, knowledge transfer, and technical assistance that aim to enable government officials as well as local actors to prepare for, and participate in, the eventual operationalization of a REDD+ mechanism. By transferring skills, knowledge, and resources, these efforts have influenced the development and implementation of policies in a number of areas relevant to REDD+ (Jodoin, 2015).

Summary

The pathways framework allows researchers, practitioners, and policy makers alike to identify and assess the distinct causal mechanisms that help to illuminate whether, when, and how, market mechanisms may exert influence on domestic policy decisions and outcomes. To illustrate the role and relevance of these multiple pathways of influence, the next section will examine the dimensions of three market mechanisms that have experienced widespread global uptake in recent years.

Forest certification, legality verification, and REDD+: Origins, dimensions, and uptake

Our two-pronged analytical framework permits us to better trace the design and influence of three globally important market mechanisms that have experienced widespread uptake around the world over the past decade: (1) forest certification, (2) timber legality verification, and (3) Reducing Emissions from Deforestation and Forest Degradation (REDD+). The review below demonstrates how the distinct ways in which these instruments draw on the four dimensions of market design, fundamentally influences the causal logic through which support, and influence, might occur.

Forest certification

Origins

Forest certification rose to global prominence as a market mechanism when environmental groups, social activists, and like-minded businesses came together to create the "Forest Stewardship Council" (FSC) in 1993 (Cashore *et al.*, 2004). The FSC's approach was relatively straightforward: develop a set of wide-ranging rules governing sustainable forest management and turn to consumers of global forest products to encourage adherence to the standards. Strategists reasoned that by drawing on carrots (price premiums) and sticks (boycotts of tropical timber and shaming campaigns for companies not undertaking certification), economic incentives might provide more enduring support and purposeful rules than international efforts that many NGOs felt amounted to "logging charters." The FSC approach also institutionalized a policy-making process that gave equal weight to environmental, social and economic factors, while creating "bottom up" working groups to develop local standards consistent with global principles and criteria (Elliott, 1999).

Originally, support for forest certification came from NGOs that saw certification as a faster means to influence forest practices than slow intergovernmental processes. Eventually forest owner and industry associations banded together to create competing certification systems with fewer prescriptive standards that were touted as more "business friendly" and were eventually housed under the global umbrella of the Program for the Endorsement of Forest Certification (PEFC) (Cashore *et al.*, 2004). This competition has resulted in highly dynamic changes to certification standards in both systems, as well as the mechanics of certification and tracking systems (Cashore *et al.*, 2006).

Instrument dimensions

Certification's main logic as a market mechanism, and what distinguishes it from legality verification or REDD+, is that it derives its authority not from the state, but from decisions made by purchasers along global supply chains about whether to purchase certified products. This means that whether producers agree to abide by the rules is related to whether the benefits of abiding by FSC or PEFC rules are higher (through price premium or market access), than the costs of compliance (meeting pre-established standards).

This distinction is important because certification systems in general, and the FSC in particular, look and act, in many ways, like traditional governments undertaking regulatory approaches. This is because standards are set, just like regulations, and firms are audited for compliance against them. Only after passing such an audit can firms benefit from the economic incentives that might accrue through such recognition. Hence, certification systems do not rely on any market logic having to do with "efficiency" or direct payments for ecosystem services. Likewise, their market logic/approach does not rely on interaction with other levels of national or local governments. If the existing level of support is strong enough to give sufficient economic incentives for compliance, firms and managers will respond irrespective of government rules. As our pathways review below emphasizes, certification systems do indeed interact with many efforts to influence public policies, and governments often undertake efforts to support or oppose the FSC. Our point here is that the institutional dimensions of certification systems do not have to interact with other levels of government for influence (Bernstein and Cashore, 2007).

Uptake

Since the early 1990s, efforts to promote responsible stewardship through certification have been mixed. On one hand, there is now considerable support for third-party certification among most commercial forestry operations in North America. There are more than 800 members of the Forest Stewardship Council responsible for over 184 million ha of certified forest in 79 countries (FSC, 2014). The areas of certified forest and forest certification are increasing the world over under the two main schemes: FSC and PEFC.

On the other hand, two challenges characterize efforts to build global uptake. First, considerable debate continues over the NGO-supported FSC and the domestic government/industry/landowner initiated PEFC programs. Second, support for certification is weakest in tropical developing countries. To date, tropical forests only account for 9.7% of the global area of FSC-certified forests and 25% of all FSC certificates (FSC, 2014). This relatively weak uptake can be traced to the enduring challenges facing tropical forest management and the rather limited economic incentives from markets in the European Union (EU) and the US compared to the costs involved in implementing certification. One exception is Bolivia, where the national government provides certified forest owners with tax benefits and exemption from government audits. This support has led to a strong certification uptake in Bolivia, in contrast to Ecuador where there is weak governmental support, although it has higher per capita income, lower corruption indices and greater access to export routes (Ebeling and Yasué, 2009).

The rise of timber legality verification appears now to be fueling forest certification uptake as more companies use certification as a means of complying with the EU Timber Regulation and US Lacey Act (Johansson, 2013). The FSC, for example, has posited that developing country markets in Latin America and Africa saw an increase in FSC-sourced materials, in order to retain access to lucrative markets in consumer countries, and to overcome the perception that most timber from "high risk" countries is illegal. This suggests greater research is required to assess the relationship between timber legality and broader forest certification systems (Cashore and Stone, 2012).

Legality verification

Origins

Timber "legality verification" is emerging as a policy instrument with which to combat illegal logging, which has been asserted to constitute some of the most egregious cases of forest degradation and deforestation (Kaimowitz, 2005), especially in tropical developing countries where biodiversity loss is a source of global concern (Tacconi, 2007). While the extent of illegal logging is very difficult to ascertain, some have estimated that up to 10% of global wood supply chains are comprised of products made from illegal wood, which has deflated world prices by 7–16% from what they would have been in the absence of illegal sources (Seneca Creek Associates, 2004). The concept of timber legality verification was endorsed in the G8 "Bali Action Plan" in 2001, which committed the world's largest global economic powers to promoting the rule of law in the forest sector.

Instrument dimensions

The ultimate aim of legality verification is to help developing-country governments enforce their own laws and policies (Cashore and Stone, 2014). The mechanism operates by tracking legal wood along global supply chains that cut across multiple jurisdictions with support generated by demand from wealthy consumer markets (Cashore and Stone, 2012). However, unlike certification, support is emerging not through consumer preferences for green products, but through trade legislation by governments such as the US, the EU and Australia, which require importers of wood products to demonstrate "due care" or "due diligence" that the products were derived from legal timber sources. In the EU, these efforts have been formalized through "voluntary partnership agreements" (VPAs) in which a producing country enters into an agreement to reduce illegal shipments to the EU.

Given that much of this economic demand is owing to changes in US and EU domestic trade policy forbidding the importation of illegal timber, the success of legality verification requires coordination by multiple levels of governance across national borders. This creates significant coordination challenges when actors along supply chains could potentially undermine efforts to support legal trade.

Uptake

The market for legally verified timber is still developing as producer countries focus on the "readiness" phase of improving their domestic systems for supply chain tracking and transparent verification processes. However, initial estimates from the US government suggest that this strategy has succeeded in increasing the market value of legal timber.[1]

At the same time, some estimate that available financing remains insufficient to meet demand and has limited the uptake of legality verification. In Indonesia alone, it is estimated that available government subsidies would only cover 5% of the US$10 million that is needed for all forest producers to obtain the government's verified legal timber certification. Other factors that limit uptake include difficulty in applying the standard to smallholders and community forests (Obidzinski *et al.*, 2014), and the inability of timber legality verification to extend influence into traditional problem areas such as land conversion to palm oil and other profitable land-uses. Part of the answer to uptake seems dependent on whether other governments will follow the lead of the EU and US in banning the importation of illegal products.

Reduced Emissions from Deforestation and Degradation (REDD+)

Origins

The origins of REDD+ can be traced back to international efforts to include developing countries within a post-2012 climate agreement. Since 2007, governments have been negotiating to establish a global mechanism within the UNFCCC to provide financial incentives to reduce carbon emissions from forestry-related sources in developing countries, which are estimated to account for 17% of greenhouse gas emissions worldwide (IPCC, 2007).

Three basic ideas have come to define the wide range of REDD+ programs, policies, and projects. First, REDD+ initiatives should finance activities that aim to increase carbon sequestration in developing-country forests by funding activities that either reduce "negative changes" or enhance "positive changes" in forest carbon stocks (Wertz-Kanounnikoff and Angelsen, 2009). Second, they should fund eligible activities on the basis of results achieved in reducing or avoiding carbon emissions or increasing carbon stocks that are measured, reported, and verified on the basis of a pre-existing baseline or reference level. Third, they should account for important social or environmental considerations beyond carbon sequestration by requiring that activities comply with social and environmental safeguards or deliver co-benefits such as poverty reduction or biodiversity preservation.

Instrument dimensions

All four dimensions of market mechanisms have been drawn on, at various stages, to promote REDD+, which itself is divided into two types of approaches: jurisdictional programs implemented by developing-country governments at the national or sub-national level as well as project-based interventions pursued at the local level. Those championing both approaches have drawn on private governing authority. Initially, private actors established certification programs enabling the proponents of REDD+ projects to validate and verify their projects and gain access to voluntary carbon markets. More recently, these same programs have established private standards and processes that apply to the jurisdictional REDD+ readiness efforts undertaken by developing-country governments.

Likewise REDD+ efforts have also drawn on "efficiency" and "payments" for behavior dimensions of market mechanisms. Decisions under the UNFCCC have specified that finance for REDD+ "may come from a wide variety of sources, public and private, bilateral and multilateral, including alternative sources" and recognize that "appropriate market-based activities" could be developed for this purpose.[2] Whether these efforts have been initiated to create innovative approaches, or simply to maintain powerful "status quo" behaviors, is hotly debated. For example, some have raised strong concerns that the integration of REDD+ into international compliance carbon markets may delay action to reduce industrial emissions in developed countries and thus undermine the overall effectiveness of the world's climate mitigation efforts (Ebeling, 2008).

Finally, REDD+, by design, captures the fourth dimension of market mechanisms in that its approach is to build a multi-level system in which global organizations and donor countries would draw on market mechanisms to influence domestic policy and practices. As such, it is envisaged that payments for REDD+ efforts could help fund national policy reforms and measures that address the drivers of deforestation, including "regulating demand for agricultural and forest products, tenure reforms, land use planning, better governance, and command and control measures" (Wertz-Kanounnikoff and Angelsen, 2009).

It is arguably for these reasons that most international efforts relating to REDD+ embrace a much larger notion of "PES-like" performance-based payments made at a jurisdictional scale rather than the direct and conditional provision of incentives at a sub-national project scale. Likewise, through multilateral funds for REDD+, activities focus on establishing the national infrastructure to receive, manage, and channel payments received for REDD+. The result is that in specific domestic contexts, such as in Indonesia, implementation of REDD+ is taking place through a complex landscape of initiatives pursued by public and private actors operating at the international, national, regional, and local levels.

Uptake

Close to 350 REDD+ projects in over 50 developing countries have been initiated by governments, international organizations, NGOs, corporations, and communities in an effort to reduce carbon emissions from forest-based sources at the local level.[3] On-the-ground, global efforts supporting the advent of REDD+ have resulted in two broad types of activities. Across Africa, Asia, Latin America, and the Caribbean, over 60 governments have initiated multi-year programs of research, capacity building, and reform to prepare for the implementation of REDD+ (known as "readiness efforts") and have begun taking action to reduce carbon emissions originating in their forests (Jodoin, 2015). Although nearly US$11 billion of public funding has been pledged for REDD+,[4] there is a significant gap between available REDD+ financing and the amount needed to ensure emissions reductions in global forests. This readiness funding is expected to gradually decline as the readiness phase ends, pilot projects are scaled up to the national level, and the voluntary and compliance carbon markets generate consistent demand for carbon offsets.

The prevalence and sustainability of existing demonstration projects has been constrained by the low international demand for carbon credits generated through REDD+ activities. In 2012, only 36% of available carbon offsets were placed with a buyer on the voluntary carbon market (Peters-Stanley *et al.*, 2013). Unless demand increases with the establishment of a compliance market for REDD+ credits, project developers will be hard pressed to unload the vast volume of forest carbon offsets in the pipeline valued at an estimated US$10.7 billion, 93% of which is from REDD projects.

For these reasons it is clear that REDD+ must be assessed for the multifaceted, changing, and evolutionary trajectory pathways through which supporters attempt to nurture these efforts. As Angelsen and McNeill remind us, one key aspect of REDD+ has been to integrate a range of actors around a collective approach to addressing related global forest challenges including, but not limited to, climate emissions (Angelsen and McNeill, 2012).

Forest certification, legality verification, and REDD+: Assessing four pathways of influence

This section applies the pathways framework to the three market mechanisms under discussion in this chapter: forest certification, REDD+, and timber legality verification. We discuss how each mechanism takes a unique route to travel all four pathways in its efforts to influence domestic forest policy. The uptake and implementation of forest certification, REDD+ and timber legality verification are illustrated in four case studies that demonstrate how market mechanisms can achieve influence through means other than the economic incentives for which they were designed.

Forest certification

Global trends

Nurturing certification through the *markets pathway* has resulted in important but incremental growth. This partly explains why REDD+, legality verification and now "no net deforestation commitments" are emerging as the latest preferred policy instruments.

However, the role of certification is not limited to the markets pathway alone. A number of efforts have been initiated under the *international rules pathway*. For example, the FSC has played an important role, along with like-minded certification programs in other arenas, in championing new global rules about product certification – largely owing to the influence of the International Social and Environmental Accreditation and Labeling (ISEAL) alliance (Bernstein and Hannah, 2008) as a counterpoint to the International Organization for Standardization (ISO), which has been criticized for being industry dominated. Certification has arguably contributed to global discussions around the regulation of global commodities, which places in historical context the EU and US roles in drawing on the *international rules pathway* to shape global preferences for tracking forest products.

Perhaps more importantly, there is no question that forest certification has influenced behaviors through traveling the *norms pathway*. For example,

FSC deliberations around old growth forest standards gave rise to the concept of "high conservation value forests" (HCVF) – which has now found its way into an array of global and domestic policy deliberations outside of certification. Indeed, so important has HCVF become that it now permeates sectors beyond forestry including palm oil, soybeans, and other agricultural commodities that have developed their own respective standards governing "High Conservation Value" arenas (Hamzah *et al.*, 2007). Similarly, a generation ago most industrial forest companies and private forest owners deemed third-party verification of their practices "inappropriate," but today, third-party verification is so entrenched that few still challenge the important role of outside auditors in assessing, and legitimizing, current practices.

Likewise forest certification has drawn upon the *direct access pathway* in a number of ways. First, by encouraging national and subnational working groups over standards development, the FSC has generated policy-oriented learning processes amongst diverse sets of stakeholders. This has been asserted to facilitate government-initiated sustainable forestry policies that otherwise would not have occurred. In fact, in Bolivia, NGOs were directly involved in revising Bolivia's forestry laws to facilitate forest certification; as a result, the government's legal criterion for sustainable forest management draws directly from FSC guidelines. In other cases, governments have used the presence of certification bodies as justification for reducing their own internal auditing processes (Cashore, 2001). In many countries, including in the developing world, scholarship has found that marginalized groups, especially indigenous communities and forest-dependent communities have gained experiences in multi-stakeholder dialogues, organizational skills, and technical training that have empowered them to access what were previously relatively closed policy-making processes (Cashore *et al.*, 2006).

Case illustration: Sweden

Certification has travelled all four pathways in influencing sustainable forestry efforts in Sweden. The *markets pathway* was a key catalyst in convincing the top five industrial forest companies in Sweden to initially support FSC certification following years of opposition (Boström, 2010; Gulbrandsen, 2005). This support followed campaigns by WWF and Greenpeace targeting UK and German purchasers of Sweden's forest products, threatening market access at a time when former Soviet bloc countries were emerging as competitors for these markets (Cashore *et al.*, 2004). It is for these reasons that the Swedish industrial forest sector was one of the largest and earliest supporters of the FSC. As of 2013, 22.5 of 40.8 million ha of Sweden's land area is covered by productive forests, of which nearly 12 million ha is certified by FSC and 11 million by PEFC (Skogsindustrierna, 2014). Some argue that this uptake has been reinforced by changes in tax law made by the Swedish government that encourage forest owners to support forest certification (Brukas and Sallnas, 2012).

Certification in Sweden also achieved influence over domestic policy by traveling the *norms pathway*; it reinforced the norms of biodiversity conservation and sustainable forestry which then pushed the Swedish government to assign equal weight to environmental goals and production goals in its national forestry policy (Cashore *et al.*, 2004). However, recent research does give us pause about just how durable these norms have been (Johansson, 2013). Several prominent environmental NGOs argue that the norms of transparency, inclusive decision-making and environmental protection championed by the FSC have not translated into Forest Management Plans (FMPs) and have largely failed to influence forest management (Sahlin, 2013).

Despite the challenges in achieving durable influence via norms, the *direct access pathway* has played a definitive role in empowering marginalized groups. In fact, FSC's Principle 9 on indigenous rights led to support of the claims of the *Sami* traditional reindeer herders, who were in dispute with Sweden's forest owners about their rights to graze reindeer over expanded areas of land. It is for these reasons that (Visseren-Hamakers and Pattberg, 2013) have concluded forest certification has had a greater impact on norms than on direct, on-the-ground environmental conditions, and on national forest policies rather than firm-level decisions.

Timber legality verification

Global trends

Timber legality verification has garnered widespread support via the *markets pathway*, due to its potential to provide market incentives in the form of access to lucrative markets in consumer countries, and the creation of a level playing field for legal timber producers in the global marketplace (Gan *et al.*, 2013). Governments support this mechanism for its potential to recapture lost tax revenue and royalties, and companies have begun to embrace supply chain tracking systems as a means of doing "good business" that streamlines production processes, improves communications with suppliers, and improves procurement practices in ways that can result in greater cost-savings (Nogueron *et al.*, 2013). However, overemphasizing the process for licensing legal timber has inadvertently favored large-scale companies over smallholders; by setting standards too high for local producers, legality verification in many cases has accidentally created barriers to entry to export markets for small-and-medium enterprises (Lesniewska and McDermott, 2014).

The market incentives provided by legality verification differ from forest certification because demand for legal timber products has not emerged owing to consumer preferences for green products. Instead, market demand has emerged through the *international rules pathway*. The US, the EU, and Australia have all enacted legislation that requires importers of wood products to proactively require that imported products are not derived from illegal sources (Cashore and Stone, 2014). Further, this trade legislation has been partially driven by the *norms pathway*, as the emergence of global norms such as "good forest governance" and "responsible procurement" has spurred action on the part of government and private sector actors alike. However, these rules have largely gained traction in exporting countries like Indonesia, which are looking to obtain or retain access to lucrative Western markets. In Brazil, where much timber is consumed domestically, domestic producers have less incentive to comply with international rules.

Timber legality verification has achieved further traction and influence through its use of the *direct access pathway*. In practice, this access takes place in negotiating and designing timber legality assurance systems, and auditing compliance and implementation. Countries which have opted to negotiate VPAs under the EU FLEGT mechanism have found that the process of negotiating VPAs can lead to the initiation of far-reaching processes of legal reform and has created an impressive array of institutional mechanisms for auditing, monitoring, and reviewing domestic timber industry operations (Overdevest and Zeitlin, 2013). Since legality verification relies on third-party auditing and supply chain tracking as a means of assuring legal compliance, the result, in many countries, has been the emergence of more actors involved in forest policy-making processes. In addition, sovereign governments are increasingly allowing non-domestic and non-governmental organizations to act as auditors for environmental compliance or as monitors of implementation (in the case of Indonesia), which creates space for third-party actors to directly take part in the governance of forest resources (Cashore and Stone, 2012). International actors also exercise influence via the *direct access pathway*; to achieve compliance with international trade legislation, forest-product-importing companies may mandate their suppliers to alter their forestry practices, or provide technology transfer, financing, and capacity building in efforts to build supply chain tracking systems (Bueno and Cashore, 2013).

Case illustration: Indonesia

The extent of illegal logging in Indonesia is widely acknowledged to be one of the highest in the world, and costs the Indonesian economy between US $1 and 5 billion per year (Seneca Creek Associates, 2004). Timber legality verification, as a mechanism for addressing illegal logging, has gained traction in Indonesia by attracting support from a broad coalition of actors motivated by increased access to global timber markets and the promise of achieving environmental goals. This coalition jointly developed Indonesia's national timber legality verification system (SVLK), which was signed into law in 2008 and is mandatory for all domestic timber product exporters.

Legality verification gained traction through all four of the pathways over the course of a decade. The Indonesian government was initially motivated by the *markets pathway*, i.e. market access to the EU. To achieve this, they entered into VPA negotiations with the EU under the *international rules pathway*, and began developing a domestic timber legality standard. The government allowed a multi-stakeholder group to have *direct access and involvement* in the standard development process, which had the intended impact of improving the credibility of SVLK. Civil society representatives were able to bring *norms* of good forest governance, transparency and accountability into the process and built provisions for third-party auditing and independent monitoring into the mechanism. This strategy appears to have paid off – the EU–Indonesian VPA was signed in 2013 and ratified by the EU parliament in 2014 – making Indonesia the first VPA country to reach this step (Yulisman, 2014). Timber exports to the EU rose by 11.8% in the first quarter immediately following ratification (Suherjoko, 2014).

Indonesia still has to address the issue of creating a domestic market for legal timber if SVLK is to affect illegal logging. Some 80% of Indonesia's timber harvest is for domestic consumption, much of which is harvested by local chainsaw operators who contribute directly to the local economy. Currently, legal timber is exported to more economically advantageous markets, such as processors in Java or provincial capitals, which means that there is almost none left for local consumption. It is clear that additional supporting mechanisms that draw upon the other pathways (certification subsidies or incentives, capacity building, anti-corruption measures) will be needed to implement SVLK effectively (Obidzinski *et al.*, 2014).

REDD+

Global trends

The *markets pathway* of REDD is constructed by governance; government regulations and policies drive market growth as they create demand from compliance with emission reduction commitments, and serve as a source of financing for project development (Peters-Stanley *et al.*, 2013). The primary set of international rules that exist for REDD+ consist of the series of decisions adopted within the UNFCCC since 2010. These include requiring the "full and effective participation of relevant stakeholders, including Indigenous Peoples and local communities".[5] The establishment of mechanisms for consultation, stakeholder engagement, and participation for REDD+ are likely to be challenging in most developing countries, in light of the poor performance of existing democratic processes and the history of mistrust that exists between government officials and local communities (Peskett and Brockhaus, 2009). On the other hand, there is some emerging evidence that they are assisting in inducing participating developing countries to adopt the laws and policies that are needed to support or implement community forestry through REDD+ (Jodoin, 2015).

In the absence of compliance markets and a binding international climate change treaty, those championing REDD+ have liberally used the *direct access pathway* to influence domestic processes. A range of multilateral, bilateral, and non-governmental actors are providing financial and technical assistance aimed at creating REDD+ "readiness" and "pilot projects." These in turn, may influence a range of domestic practices. For example, project-scale REDD+ activities may have important implications for promoting participatory rights and community forestry. Due to the importance of aid funding in the landscape of REDD+ activities, many early projects have, in the name of a "pro-poor" approach, sought to support and empower forest-dependent communities undertaking or committed to sustainable forest management practices (Lawlor *et al.*, 2013). Perhaps the most important effect of the *direct access pathway* has been the support for the emergence of new actors in domestic policy-making processes, as international experts, bilateral fund representatives, indigenous peoples and forest-dependent communities have been more directly involved in developing REDD+ strategies than in previous forest policy-making efforts.

REDD+ projects may also generate lessons about the role and importance of different approaches such as community forestry within national REDD+ readiness efforts. REDD+ pilot projects may provide policy-makers with information about the costs, feasibility, and effectiveness of various project-level interventions (Jagger *et al.*, 2009). Field interviews conducted by one of this chapter's authors have found that certification programs for REDD+ projects such as the Climate, Community and Biodiversity (CCB) Alliance standard have created standards and processes that some assert have contributed to the effectiveness of conservation projects in ways that did not exist prior to REDD+ (Jodoin, 2015).

Case illustration: Indonesia

Following application of REDD+ to Indonesia, the government there announced plans to reduce emissions by as much as 41% by 2020. While the precise causal influence of REDD+ on this goal is difficult to assess, we do know that Indonesia's *efforts to reduce emissions* have spurred forest governance reforms that may achieve outcomes far beyond *emissions reductions alone*. In other words, REDD+ seems to have catalyzed shifts in norms and business-as-usual practices in forest governance, especially with regard to the rights of indigenous peoples and forest-dependent communities.

These emerging norms have been formalized within the *international rules pathway*, which paved the way for uptake within Indonesia's national policy processes. The Indonesian government has largely accepted the legitimacy and fairness of existing rules regarding REDD+, and has complied with them in developing a national REDD+ strategy and systems for addressing safeguards.

Several scholars reason that in Indonesia market incentives alone will be insufficient, because REDD+ financing is unable to offset the economic benefits from converting land to more profitable uses, such as oil palm, mining, or agriculture. Where the *markets pathway* has succeeded is in pushing forward standards for certifying carbon credits generated through REDD+ activities. These standards and processes have enhanced the accountability and effectiveness of conservation projects in ways that did not exist prior to REDD+.

The case of Indonesia demonstrates how pursuing the *markets pathway*, in preparing to enact REDD+ activities, seems to have influenced forestry policy-making. In Indonesia, the UN-REDD Program undertook multiple activities to support the development of social safeguards and instruments for REDD+, especially with respect to the free, prior, and informed consent (FPIC) of Indigenous Peoples and local communities. This most notably included pilot activities aimed at testing FPIC protocols in two villages in the province of Central Sulawesi (UN-REDD, 2012a) and the organization, with the National Forestry Council of Indonesia, of a multi-stakeholder process that resulted in a set of policy recommendations on how FPIC could be implemented in Indonesia's REDD+ laws and institutions (UN-REDD, 2012b). While the impact of REDD+ continues to be highly dynamic, what is clear is that in addition to harnessing market forces, REDD+ efforts have created new norms that challenge "business-as-usual practices," and through the *international rules pathway*, have helped foster enhanced multi-stakeholder practices.

Summary

This section examined how forest certification, REDD+, and timber legality verification have influenced forestry policy, using case studies from Sweden and Indonesia. What is clear is that these three mechanisms have not only traveled the *markets pathway* by efforts to nurture "low carbon" economic signals but they have also traveled the *norms, international rules*, and *direct access pathways* in important ways.

Integrating policy dimensions and pathways: Strategic implications for supporting and nurturing market mechanisms

The four pathways of influence and the four dimensions of market instruments allow us to generate several insights about the conditions through which market mechanisms might promote sustainable forestry practices.

First, our review suggests that *market mechanisms that draw on private authority are likely to be insufficient for addressing deforestation, degradation, and climate challenges.* Twenty years of research on certification systems, and more recent efforts on voluntary carbon markets in the forest sector, have revealed important and growing levels of support on the one hand, but relatively modest market shares on the other hand. Since the promise of supply chain governance rests on fully institutionalized supply chains, it is safe to conclude that such status, if it is to occur, is years or decades away. This means that by themselves, market mechanisms seem poorly designed to fully address urgent issues such as the climate crisis and deforestation. At the same time, *private authority has had significant, yet largely poorly studied, effects on international rules, norms, and direct access pathways.* Supporters of certification, and the FSC in particular, argue they have had profound but difficult-to-measure impacts in shifting power dynamics, problem definitions, and domestic policy processes. Likewise, under REDD+, the voluntary carbon market has created valuable learning opportunities about the design and settings of carbon instrument choices, which may facilitate and render more efficient government efforts to follow (Bozzi *et al.*, 2012). When designed well, private authority may even help to "fill the gaps" in international agreements that can create synergistic legitimacy granting interactions across public and private authority (Levin *et al.*, 2009).

Second, practitioners must assess not only the potential for traveling all four pathways, they must also identify, and undertake strategies consistent with, the multiple causal steps that may lead to meaningful "on-the-ground" results. Such complex pathways might include supporting private authority for non-direct impacts. Some argue, for example, that green building standards certification systems will have limited impact on building decisions given their niche status, but they could play a significant role in shaping learning about green building practices, which might then diffuse to powerful actors who set municipal building codes (Steering Committee, 2012).

Likewise, PES efforts must not only be assessed for their direct economic incentives, but as to whether they might help "tip the scales" domestically toward conservation and reducing deforestation. This turns attention to assessing whether the price of REDD+ forest carbon offsets might eventually be "stacked" with other economic benefits, such as FSC-certified timber, or other significant social and environmental co-benefits. Given that many REDD+ projects encompass the production of agricultural commodities certified under various eco-labeling schemes, such as coffee, soybeans, and nuts, the added benefit of a price premium for carbon needs to be assessed. Such a premium might potentially increase the financial viability of REDD+ projects and forest certification efforts. Since deforestation and forest governance challenges are most often found in fragile states, or countries with weak governance systems, it seems plausible to conclude that relying on financial incentives alone, or any single instrument or mechanism, will be insufficient to overcome these seemingly intractable challenges (Aquino and Guay, 2013). Rather than jettison such approaches, a more fruitful direction seems to lie in carefully building problem-oriented "policy baskets" (Howlett and Rayner, 2007) that draw on, but do not isolate, market mechanisms. For example, in countries like Brazil where enforcement of national laws remains weak, REDD+ will also require strong national command and control regulation to underpin its implementation (Taylor *et al.*, 2012).

Efforts to build private authority have created communities across environmental, social, and economic interests that focus on standards development and institutional design. These communities can be highly important for creating legitimacy (Bernstein and Cashore, 2007) and trust (McDermott, 2012), but they can be destructive if building the mechanism trumps problem solving. For example, Cohen *et al.* (1972) found that policy makers and practitioners in an organization tend to start with their preferred mechanism and look for problems to apply it to, rather than starting with the problem and casting around for solutions. It is important that practitioners not be blinded by their instrument choice and instead focus on the ultimate objective being pursued.

Third, the direct access pathway needs greater attention, because, if executed thoughtfully and purposefully, it can empower marginalized groups. Our review above found that both legality verification and REDD+ are designed to influence domestic policy and behavior. This means paying attention to the role of norms in shaping problem definitions, the broader social and environmental values that could be reduced or enhanced, and the capacity of stakeholders to engage in policy deliberations (McDermott, 2014b). Most important, the largest impact of these market mechanisms may not be on carbon reductions, but on the emergence of transnational law that embeds norms of human rights, livelihoods, and indigenous community resources across multiple levels. In other words, far from reinforcing neo-liberal norms, market mechanisms may actually undermine them, producing a global citizenry that, united through supply chains, broadens the nature of communities and the problems they seek to address.

Clearly, the decision whether to draw on market mechanisms is not straightforward. For those practitioners and scholars who focus only on instrument design or implementation, the elephant in the room is that the *markets pathway* alone may actually fail to efficiently harness market forces or deliver the economic incentives needed to achieve the desired goal (reduced forest loss). If the mechanism is being championed for ideological reasons, with no discernible "causal logic" in how it might address a specified "on-the-ground problem," then practitioners must ask whether powerful interests are limiting options, rather than finding creative solutions. The implication of these findings is important for how practitioners evaluate, and develop, new ideas for improving REDD+, forest certification, or legality verification efforts, as well as new innovations, such as recent initiatives to address deforestation through "no net deforestation by 2020" commitments.

To address this elephant, it is crucial that scholars and practitioners better understand the mechanisms and interactions that can create countervailing efforts and synergies, while minimizing redundancy, incoherence, or undermining other mechanisms (Hogl, 2002). Better understanding of these influence dynamics is needed to create strategies to maximize these positive outcomes while minimizing negative outcomes. The framework offered here, which distinguishes key dimensions of market mechanisms, and four pathways through which these mechanisms achieve influence, offers a comprehensive approach to carefully assessing and identifying the potential, pitfalls, and opportunities of market mechanisms that clearly reach far beyond the market itself.

Notes

1 The US's trade balance of forest products went from a US$20.3 billion deficit to a US$600 million surplus between 2006 and 2010, the same period during which the Lacey Act was amended (http://democrats.naturalresources.house.gov/press-release/defazio-fights-protect-american-jobs-curb-illegal-timer-trade (accessed July 16, 2014).
2 Decision 2/CP.17, Outcome of the work of the Ad Hoc Working Group on Long-term Cooperative Action under the Convention, FCCC/CP/2011/9/Add.1 (March 15, 2012) at para. 65–66.
3 See CIFOR, "Global database of REDD+ and other forest carbon projects Interactive map," www.forestsclimatechange.org/redd-map (accessed June 10, 2014).
4 According to the most recent available data from the voluntary REDD+ Database 2014. "How much financing has been reported for REDD+?," www.fao.org/forestry/vrd/#graphs_and_stats (accessed July 5, 2014).
5 Decision 1/CP.16, para. 72.

References

Angelsen, A. and McNeill, D. (2012) 'The evolution of REDD+', in A. Angelsen, M. Brockhaus, W.D. Sunderlin and L.V. Verchot (eds) *Analysing REDD+: Challenges and Choice*, Center for International Forestry Research, Bogor, Indonesia, pp. 31–49.

Aquino, A. and Guay, B. (2013) 'Implementing REDD in the Democratic Republic of Congo: an analysis of the emerging national REDD governance structure', *Forest Policy and Economics*, 36, pp. 71–79.

Auld, G. and Cashore, B. (2012) 'Appendix F, Forestry Review', in Steering Committee of the State-of-Knowledge Assessment of Standards and Certification. Toward Sustainability: The Roles and Limitations of Certification, Resolve, Washington, DC.

Bernstein, S. (2001) *The Compromise of Liberal Environmentalism*, Columbia University Press, New York.

Bernstein, S. and Cashore, B. (2000) 'Globalization, four paths of internationalization and domestic policy change: the case of eco-forestry in British Columbia, Canada', *Canadian Journal of Political Science*, 33(1), pp. 67–99.

Bernstein, S. and Cashore, B. (2007) *Can Non-State Global Governance Be Legitimate? Overcoming the Conundrum of Market-based Authority*. Yale University, New Haven, CT.

Bernstein, S. and Hannah, E. (2008) 'Non-state global standard setting and the WTO: legitimacy and the need for regulatory space', *Journal of International Economic Law*, 11(3), pp. 575–608.

Betsill, M.M. and Hoffmann, M.J. (2009) 'Constructing "Cap and Trade": The Evolution of Emissions Trading Systems for Greenhouse Gases', Paper prepared for the 2009 Open Meeting of the Human Dimensions of Global Environmental Change Research Community, Bonn, Germany.

Boström, M. (2010) *The Evidence Base for Environmental and Socioeconomic Impacts of 'Sustainable' Certification*. Stockholm Centre for Organizational Research (SCORE), Stockholm, Sweden.

Bozzi, L., Cashore, B., Levin, K. and McDermott, C. (2012) 'The role of private voluntary climate programs affecting forests: assessing their direct and intersecting effects', in K. Ronit (ed.) *Business and Climate Policy: The Potentials and Pitfalls of Private Voluntary Programs*, United Nations University Press.

Brukas, V. and Sallnas, O. (2012) 'Forest management plan as a policy instrument: carrot, stick or sermon?' *Land Use Policy*, 29, pp. 605–613.

Bueno, G. and Cashore, B. (2013) 'Can legality verification combat illegal logging in Brazil? Strategic insights for policy makers and advocates', in IUFRO Task Force on Forest Governance, *Issues and Options Briefs*, International Union of Forest Organizations.

Cashore, B. (2001) *In Search of Sustainability: British Columbia Forest Policy in the 1990s*, UBC Press, Vancouver, Canada.

Cashore, B. and Stone, M.W. (2012) 'Can legality verification rescue global forest governance? Analyzing the potential of public and private policy intersection to ameliorate forest challenges in Southeast Asia', *Forest Policy and Economics*, 18, pp. 13–22.

Cashore, B. and Stone, M.W. (2014) 'Does California need Delaware? Explaining Indonesian, Chinese, and United States support for legality compliance of internationally traded products', *Regulation and Governance*, 8, pp. 49–73.

Cashore, B., Auld, G. and Newsom, D. (2004) *Governing Through Markets: Forest Certification and the Emergence of Non-State Authority*, Yale University Press, New Haven, CT.

Cashore, B., Gale, F., Meidinger, E. and Newsom, D. (eds) (2006) *Confronting Sustainability: Forest Certification in Developing and Transitioning Societies*, Yale School of Forestry and Environmental Studies, New Haven, CT.

Cohen, M.D., March, J.G. and Olsen, J.P. (1972) 'A garbage can model of organizational choice', *Administrative Science Quarterly*, 17(1), pp. 1–25.

Ebeling, J. (2008) 'Risks and criticisms of forestry-based climate change mitigation and carbon trading' in C. Streck, R. O'Sullivan, T. Janson-Smith and R. Tarasofsky, (eds) *Climate Change and Forests: Emerging Policy and Market Opportunities*, Chatham House, London.

Ebeling, J. and Yasué, M. (2009) 'The effectiveness of market-based conservation in the tropics: forest certification in Ecuador and Bolivia', *Journal of Environmental Management*, 90(2), pp. 1145–1153.

Elliott, C. (1999) *Forest Certification: Analysis from a Policy Network Perspective*. PhD thesis, Departement de genie rural, Ecole Polytechnique Federale de Lausanne, Lausanne, Switzerland.

Elliott, C. (2005) 'From the tropical timber boycott to forest certification' in D. Burger, J. Hess and B. Lang (eds) *Forest Certification: An Innovative Instrument in the Service of Sustainable Development?* GTZ, Programme Office for Social and Ecological Standards, Eschborn, Germany, pp. 79–90.

FSC (Forest Stewardship Council) (2014) *Global FSC Certificates: Type and Distribution October 2014*. Forest Stewardship Council, Bonn, Germany.

Gan, J., Cashore, B. and Stone, M.W. (2013) 'Impacts of the Lacey Act Amendment and the Voluntary Partnership Agreements on illegal logging: implications for global forest governance', *Journal of Natural Resources Policy Research*, 5(4), pp. 209–226.

Grieg-Gran, M., Porras, I. and Wunder, S. (2005) 'How can market mechanisms for forest environmental services help the poor? Preliminary lessons from Latin America', *World Development*, 9(33), pp. 1511–1527.

Gulbrandsen, L.H. (2005) 'The effectiveness of non-state governance schemes: a comparative study of forest certification in Norway and Sweden', *International Environmental Agreements: Politics, Law and Economics*, 5(2), pp. 125–149.

Gunningham, N., Grabosky, P.N. and Sinclair, D. (eds) (1998) *Smart regulation: Designing Environmental Policy, Oxford Socio-legal Studies*, Clarendon Press and Oxford University Press, Oxford and New York.

Hamzah, K.A., Nik, A.R., Efransjah, C., Hassan, H. and Jarvie, J. (2007) 'The assessment of High Conservation Value Forest (HCVF) of the South-East Pahang peat swamp forest, Malaysia – a case study', *Malaysian Forester*, 70(2), pp. 133–143.

Hogl, K. (2002) 'Reflections on inter-sectoral coordination in national forest programmes', EFI Proceedings, Cross-sectoral policy impacts on forests.

Howlett, M. and Rayner, J. (2007) 'Design principles for policy mixes: cohesion and coherence in "new governance arrangements"', *Policy and Society*, 4(26), pp. 1–18.

IPCC (2007) *Intergovernmental Panel on Climate Change – IPCC, Fourth Assessment Report*, Cambridge University Press, Cambridge, UK.

Jagger, P., Stibniati, A., Pattanayak, S.K., Sills, E. and Sunderlin, W.D. (2009) 'Learning while doing: evaluating impacts of REDD+ projects' in A. Angelsen, M. Brockhaus, W.D. Sunderlin and L.V. Verchot (eds) *Analysing REDD+: Challenges and Choice*, CIFOR, Bogor, Indonesia, pp. 281–292.

Jodoin, S. (2015) *Forests, Carbon and Rights: The Transnational Legal Process for REDD+ and the Rights of Indigenous Peoples and Local Communities*. PhD Dissertation, Yale University, Graduate School of Arts and Sciences.

Johansson, J. (2013) 'Constructing and contesting the legitimacy of private forest governance: the case of forest certification in Sweden', Department of Political Science Research Report, Umea University, Umea, Sweden.

Joskow, P.L. and Schmalensee, R. (1998) 'The political economy of market-based environmental policy: the U.S. acid rain program', *The Journal of Law and Economics*, 41(1), pp. 37–84.

Kaimowitz, D. (2005) 'Illegal logging: causes and consequences', Paper read at the Forests Dialogue on Illegal Logging, Hong Kong.

Lawlor, K., Madeira, E., Blockhus, J. and Ganz, D. (2013) 'Community participation and benefits in REDD+: a review of initial outcomes and lessons', *Forests*, 2(4), pp. 296–318.

Lesniewska, F. and McDermott, C.L. (2014) 'FLEGT VPAs: laying a pathway to sustainability via legality lessons from Ghana and Indonesia', *Forest Policy and Economics*, 48, 16–23.

Levin, K., Cashore, B. and Koppell, J. (2009) 'Can non-state certification systems bolster state-centered efforts to promote sustainable development through the Clean Development Mechanism (CDM)?' *Wake Forest Law Review*, 44, pp. 777–798.

Levin, K., Cashore, B., Bernstein, S. and Auld, G. (2012) 'Overcoming the tragedy of super wicked problems: constraining our future selves to ameliorate global climate change', *Policy Sciences*, 45(2), pp. 123–152.

Lewis, D.J. (2001) 'Easements and conservation policy in the North Maine woods', *Maine Policy Review*, 10(1), pp. 24–36.

McDermott, C.L. (2012) 'Trust, legitimacy and power in forest certification: a case study of the FSC in British Columbia', *Geoforum*, 43(3), pp. 634–644.

McDermott, C.L. (2014a) 'Forest certification and legality initiatives in the Brazilian Amazon: lessons for effective and equitable forest governance', *Forest Policy and Economics*, 50, pp. 134–142.

McDermott, C.L. (2014b) 'REDDuced: from sustainability to legality to units of carbon – the search for common interests in international forest governance', *Environmental Science and Policy*, 35, pp. 12–19.

McGinnis, M.D. (ed.) (1999) *Polycentric Governance and Development: Readings from the Workshop in Political Theory and Policy Analysis*, University of Michigan Press, Ann Arbor, MI.

Nogueron, R., Hess, K. and Refkin, D. (2013) 'Untangling the paper chain: how Staples is managing transparency with suppliers', World Resources Institute Issue Brief, World Resources Institute, Washington, DC.

Obidzinski, K., Dermawan, A., Andrianto, A., Komarudin, H. and Hernawan, D. (2014) 'The timber legality verification system and the voluntary partnership agreement (VPA) in Indonesia: challenges for the small-scale forestry sector', *Forest Policy and Economics*, 48: 24–32.

Overdevest, C. and Zeitlin, J. (2013) 'Constructing a transnational timber legality assurance regime: architecture, accomplishments, challenges', *Forest Policy and Economics*, 48, 6–15.

Perman, R. (2003) *Natural Resource and Environmental Economics*, Pearson Education, Harlow, UK.

Peskett, L. and Brockhaus, M. (2009) 'When REDD+ goes national: a review of realities, opportunities and challenges' in A. Angelsen, M. Brockhaus, W.D. Sunderlin and L.V. Verchot (eds) *Analysing REDD+: Challenges and Choice*, CIFOR, Bogor, Indonesia, pp. 25–41

Peters-Stanley, M., Gonzalez, G. and Yin, D. (2013) *Covering New Ground: State of the Forest Carbon Markets 2013*, Forest Trends Ecosystem Marketplace, Washington, DC.

Sahlin, M. (2013) *Credibility at Stake: How FSC Sweden Fails to Safeguard Forest Biodiversity*. Swedish Society for Nature Conservation, Stockholm, Sweden.

Sasser, E.N., Prakash, A., Cashore, B. and Auld, G. (2006) 'Direct targeting as an NGO political strategy: examining private authority regimes in the forestry sector', *Business and Politics*, 8(3), pp. 1–32.

Seneca Creek Associates, LLC (2004) *'Illegal' Logging and Global Wood Markets: The Competitive Impacts on the U.S. Wood Products Industry*, American Forests and Paper Association, Poolesville, Maryland.

Skogsindustrierna (2014) *The Swedish Forest Industries Facts and Figures 2013*, Swedish Forest Industries Federation, Stockholm, Sweden.

Steering Committee (Steering Committee of the State-of-Knowledge Assessment of Standards and Certification) (2012) *Toward Sustainability: The Roles and Limitations of Certification*, Resolve, Washington, DC.

Suherjoko (2014) 'EU ratification helps push timber exports in first quarter', *The Jakarta Post*. 15 August 2014. Available from: www.thejakartapost.com/news/2014/08/15/eu-ratification-helps-push-timber-exports-first-semester.html (Accessed 17 August 2014).

Tacconi, L. (ed.) (2007) *Illegal Logging: Law Enforcement, Livelihoods and the Timber Trade*, Earthscan, London.

Taylor, C., Pollard, S., Rocks, S. and Angus, A. (2012) 'Selecting policy instruments for better environmental regulation: a critique and future research agenda', *Environmental Policy and Governance*, 22(4), pp. 268–292.

UN-REDD (2012a) *Free, Prior and Informed Consent for REDD+ in the Asia-Pacific Region: Lessons Learned*, UN-REDD Programme, Geneva, Switzerland.

UN-REDD (2012b) *Policy Recommendation: Free, Prior and Informed Consent (FPIC) Instrument for Indigenous Communities and/or Local Communities Who Will Be Affected by REDD+ Activities*, UN-REDD Programme Indonesia and the National Forestry Council, Jakarta, Indonesia.

Visseren-Hamakers, I. and Pattberg, P. (2013) 'We can't see the forest for the trees: the environmental impact of global forest certification is unknown', *GAIA*, 22(1), pp. 25–28.

Wertz-Kanounnikoff, S. and Angelsen, A. (2009) 'Global and national REDD+ architecture: linking institutions and actions' in A. Angelsen (ed.) *Realising REDD+, National Strategy and Policy Options*, CIFOR, Bogor, Indonesia, pp. 13–24.

Yulisman, L. (2014) 'Indonesia aims long-term timber export growth in the EU', *The Jakarta Post*. 1 March 2014. Available from: www.thejakartapost.com/news/2014/03/01/indonesia-aims-long-term-timber-export-growth-eu.html (Accessed 17 August 2014).

5 On corporate responsibility

Anne Toppinen, Katja Lähtinen and Jani Holopainen

Introduction

In the 1970s the song "Down by the river" became an iconic expression in popular music of environmental pollution problems, which at that time in northern Europe were to a large extent caused by pulp and paper factories. In Finland, a Finnish version of the song, performed by Kirka, reached third place in the singles charts in summer 1973, and even most children of the time knew how to sing its chorus. Public concern over the activities of the pulp and paper sector ushered in more stringent environmental regulation in Scandinavia and later elsewhere in Europe. The forest industry made large-scale investments to curb atmospheric and aquatic emissions during the next decades; despite substantial growth in production, current emission levels are far below their compliance levels and are marginal in comparison to the situation 30 years ago. Thus, today it is quite possible in the pulp and paper countries of northern Europe to take a trip down by the river without having health concerns!

The forest industry has always been an environmentally sensitive sector, due to its heavy reliance on natural resources and the profound impact it can have on vital ecosystems, and this has not changed. For example, according to Kozak (2013, 432), "*there is no force – be it climate change, pests, disease, fire, poverty, and so on – that has as big an impact on the current and future states of our forests as business has.*" Furthermore, the understanding and practice of corporate responsibility in the forest industry depends on the context, particularly in terms of the region and the size of the company (Vidal and Kozak 2008). Due to large-scale global trade in forest products, environmental and socio-cultural impacts are also often global in scope, regardless of whether the company only operates domestically (Lähtinen and Myllyviita 2014; Mayer *et al.* 2005). Furthermore, in the past two decades, internationalization of the forest industry has led to expansion of production in emerging and developing countries in the global South, leading to renewed challenges to the legitimacy of forestry operations (e.g. Mikkilä and Toppinen 2008). Therefore, based on the literature, what can we conclude about the current state and future development of corporate responsibility in this field?

In this chapter we first review the background of corporate (social) responsibility (hereafter CR) and then introduce some of the more widely used measurement tools and administrative instruments used to measure and communicate sustainability in global forest industries. Second, we discuss the role of CR in global forest industries from the perspective of different stakeholder groups. Third, we focus on a set of CR initiatives and activities including corporate sustainability disclosure, usage and adoption of CR management systems (e.g. ISO 14001 on environmental impacts or SA8000 on social impacts), the use of product labels and declarations (e.g. Programme for the Endorsement of Forest Certification, PEFC, or Environmental Product Declarations, EPDs), initiatives aimed at carbon offsetting and offering guidance to investors (e.g. Carbon Disclosure Project, CPD), and the inclusion of forestry companies in various sustainable/ethical investment indices. Fourth, we discuss the success and limitations of CR in terms of how well CR management and existing reporting systems reach stakeholders. The chapter ends with concluding remarks on the future challenges of CR in the forest industry.

What is corporate responsibility and why should we care about it?

The concept of corporate responsibility is multifaceted and often used in conjunction with such terms as "corporate social responsibility," "corporate sustainability," "corporate citizenship," "corporate social initiative," "corporate social responsiveness," or as a synonym for other concepts such as the "triple-bottom line" (economic, environmental, and social) and "the three Ps" (profits, planet, and people) (see e.g. Dahlsrud 2008 and Crane *et al.* 2013; for recent reviews of CR literature, see e.g. Alguinis and Glavas 2012 or Malik 2014). In this chapter we adopt a definition commonly used by the European Commission (2001), according to which CR requires that "*companies integrate social and environmental concerns in their business operations and in their interaction with their stakeholders on a voluntary basis.*" This implies corporate social and environmental behavior which goes beyond the legal (regulatory) requirements in the relevant markets. More recently, the European Commission has defined CR in simpler (but surely no less demanding) terms as the "*responsibility of enterprises for their impacts on society.*" This calls for the establishment of processes to integrate social, environmental, ethical, human rights and consumer concerns into business operations and core strategies in close cooperation with stakeholders (European Commission 2011).

According to Kurucz *et al.* (2008), companies' motivation for engaging in CR can be of four types: to reduce costs and operational risks, to achieve a competitive advantage, to improve company reputation and legitimacy, and to integrate stakeholder interests to create value on multiple fronts (i.e. synergistic value creation). Promoting a corporate culture of sustainability has the potential to achieve competitive advantage and improve performance (Eccles *et al.* 2011) via the development of valuable, rare and non-imitable organizational resources and capabilities (Barney 1986), and the integration of ecological aspects of the business environment into decision-making processes (see e.g. Hart 1995; Litz 1996; Branco and Rodrigues 2006). Building CR systemically into corporate strategy may thus help companies align themselves with the interests of stakeholders and society at large in order to create shared value (Galbreath 2009; Porter and Kramer 2006, 2011).

Among competing theories of CR, Freeman's (1984) stakeholder theory has most frequently been used as a framework for analyzing interactions between companies and society. According to Freeman (1984), stakeholders are understood as groups "*who influence or are being influenced by the actions of an organization.*" In order to improve the descriptive power of CR, Clarkson (1995) categorized stakeholder groups into primary stakeholders (those whose continuing participation is essential to the company's survival, i.e. shareholders, employees, customers, suppliers and the government) and secondary stakeholders (who do not directly interact with the company, but who are otherwise affected by it, such as communities, civil society organizations, competitors or the media). A gradual path of increased stakeholder involvement has typically included a transition through the following three phases: stakeholder identification, stakeholder management and finally stakeholder engagement (e.g. Manetti 2011). In practice, stakeholder engagement is an integral part of CR management that is implemented via management systems which also steer sustainability investments. Sustainability investments include finding synergies between different business processes through using byproducts, recycled materials and waste, reducing costs, increasing added value and managing business risks, while stakeholder engagement includes dialogue with and integration of different stakeholders (Robèrt *et al.* 2002).

An integral part of CR management systems is the translation of the impacts caused during the product life-cycle into concrete measurable data, and the formulation of recommendations that can be used as guiding principles in managerial decision-making in the global forest industries (von Geibler *et al.* 2010). In addition, when defining the scope and focus of measurements, stakeholder consultation is an integral part of CR management systems (Rasche and Esser 2006). As a result of the integration of "responsibility" into companies' strategic decision-making and by including stakeholders in value-creation processes, CR reporting strongly promotes engagement in CR and helps provide societal legitimacy (e.g. Kurucz *et al.* 2008). Nevertheless, while there is a general perception of why firms might engage in CR, there are few in-depth studies into the motivation for various stakeholder groups' involvement in CR practices.

CR management and reporting in global forest industries

Ideally, CR management should consider the economic, environmental and social impacts of all phases of the product life-cycle, from resource extraction, production and consumption to recycling and disposal after use (Table 5.1).

Table 5.1 Sustainability impacts with the potential for strategic synergic value creation in forest industry business processes, from resource extraction to disposal

Sustainability impacts	*Resource extraction*	*Production*	*Consumption*	*Recycling*	*Disposal*
Economic	Enhancing regional incomes and infrastructures	Developing new products manufactured with fewer inputs and/or comprising higher added value	Providing repair systems and/or instructions for repairing/replacing worn parts	Utilizing secondary raw materials and substitution of primary raw materials to decrease costs and/or increase added value	Increasing the utilization of waste as a valuable asset to increase efficient use of resources and energy
Environmental	Preventing forest land-use change and forest degradation	Reducing greenhouse gas emissions and increasing resource and energy efficiency	Decreasing atmospheric and aquatic emissions	Improving product design for easy recycling	Decreasing the amount of waste and increasing the separability of waste at end of life
Social	Fostering local participation	Decreasing the usage of chemicals in processes with negative impacts on human health	Enhancing the fitness for purpose in use	Providing information for recycling and/or product disassembly	Providing information for disposal and collection of worn products

Adapted from Cobut *et al.* (2013)

In distinction from traditional performance management systems, which primarily focus on economic factors, CR management systems attempt to control the economic, environmental and social impacts of business processes, and the company's value chain is extended to include the life-cycle phases of recycling and disposal (e.g. Skaar and Magerholm Fet 2012).

By extending the forest sector value chain, novel approaches can be found to such things as new materials, improved product designs and creating synergistic value with different stakeholders (e.g. von Geibler *et al.* 2010; Lähtinen *et al.* 2014). In addition, comprehensive CR management also allows the evaluation of trade-offs in sustainability impacts between business choices at different product life-cycle phases and the potential rebound effects of efficiency gains in specific parts of a company's processes (Robèrt *et al.* 2002). Thus, in terms of, say, the business processes of forest companies, integrating life-cycle thinking, which includes the notions of industrial ecology and industrial symbiosis, into CR management allows those processes to be seen as part of larger economic (e.g. business clusters), environmental (e.g. ecosystem services) and social (e.g. local communities) systems (e.g. Korhonen *et al.* 2005).

The use of measurement tools to assess the sustainability impacts of companies' business processes is a fundamental part of CR management systems. Standardized and legally binding methods for gathering financial statement information and implementing financial statement analysis can be easily employed to acquire narrowly focused information on companies' economic performance and sustainability (e.g. liquidity, turnover, growth and return on investment) (e.g. Lähtinen and Toppinen 2008). In contrast, gaining holistic and comparable information on environmental and social sustainability impacts is more challenging. Environmental Life-Cycle Assessment (ELCA) and Social Life-Cycle Assessment (SLCA) methods can be employed to acquire general information for measuring the sustainability impacts of comparable products and/or determining "hot spots," i.e. aspects of product life-cycles with the most critical environmental impacts (e.g. ozone formation, ecotoxity and effects on biodiversity) and social impacts (e.g. workers' health, safe and healthy living conditions for local communities and prevention of corruption in society) (e.g. Benoît *et al.* 2010; Myllyviita *et al.* 2013).

In comparison to SLCA, the methodologies of ELCA are currently well standardized, as a result of strong development since the 1980s (e.g. Finnveden *et al.* 2009). In contrast, for the less theoretically developed SLCA, development of assessment methodologies is still ongoing (e.g. Benoît *et al.* 2010). In the case of forest industries, ELCA has been recently used, among other things, to assess the ecological impacts of forest-based biofuel production (Myllyviita *et al.* 2012) and forest sector operations in general (Gonzaléz-Garcia *et al.* 2014). Moreover, in a comprehensive literature review, Cobut *et al.* (2013) studied the linkages between aspects of life-cycle assessments and the environmental labelling of wood products used in construction. In the context of the forest industries, SLCA has only been applied to theoretical and methodological considerations (Macombe *et al.* 2013). In spite of the less established status of SLCA as a practical sustainability assessment methodology, integration of not only the environmental impacts, but also the social life-cycle effects of companies is expected to be increasingly crucial for future CR management (e.g. Galbreath 2009).

CR reporting systems are administrative decision-making tools within CR management systems that are used to control, organize and improve the economic, environmental and social sustainability of business processes (see e.g. Lozano and Huisingh 2011). According to Niskala *et al.* (2013), pivotal examples of international CR guidelines and/or standards employed in CR reporting systems (henceforth referred to as CR standards) are *Guidelines for Multinational Enterprises* (Organisation for Economic Cooperation and Development, OECD 2011); *Global Compact* (GRI and UN 2013); *AA1000* (AccountAbility 2012); *ISO26000* and *ISO14001* (International Organization for Standardization, ISO 2010, 2004); *GRI Guidelines* (Global Reporting Initiative, GRI 2011); *Environmental Management Auditing Scheme, EMAS* (European Union, EU 2009); *Occupational Health and Safety Assessment Series, OHSAS* (British Standard Institution, BSI 2007) and *Social Accountability 8000, SA8000* (Social Accountability International, SAI 2008).

As illustrated in Figure 5.1, the existing CR standards provide, to varying degrees, information on CR management, CR measurement and the economic, environmental and social impacts of business processes. What is noteworthy, however, is the lack of standardization in the domain of stakeholder communication. Nevertheless, Panwar and Hansen's (2007) study of the applicability of CR standards in the US and Indian forest products

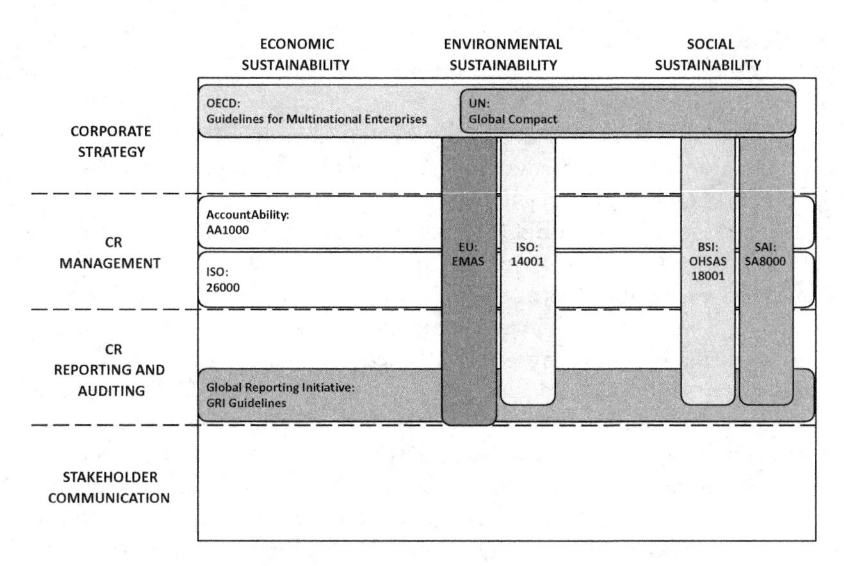

Figure 5.1 CR standards for corporate-level assessment of economic, environmental and/or social sustainability (modified from Niskala *et al.* 2013)

industries indicates an increasing impetus towards the creation of a global, internationally accepted CR standard.

CR reporting systems support management by providing a conceptual framework, i.e. principles, criteria and indicators for sustainability assessments. The more detailed the sustainability assessments, the more transparent and comparable are their results. Principles are "a fundamental truth" used as the basis for reasoning or action in sustainability assessments (e.g. "Businesses should support and respect the protection of internationally proclaimed human rights"), while criteria are concrete explanations of principles without being direct measures of performance (e.g. "Personnel shall have the right to leave the workplace premises after completing the standard workday, and be free to terminate their employment provided that they give reasonable notice to their employer") (Mendoza and Prabhu 2000). At the most detailed level of assessment, indicators (see e.g. GRI 2011) are employed as concrete measures of business processes intended to condense large amounts of information into a manageable format (e.g. Bossel 1996). In addition, for some CR reporting systems, auditing schemes exist for voluntary or mandatory certification by independent third-parties to assure the standards of companies' CR reporting.

In short, although there is already compatibility between some CR standards (e.g. GRI and UN 2013) and development work continues apace (e.g. GRI and ISO 26000), there are still notable differences between various CR reporting systems which affect, for example, the implementation costs of sustainability management practices within different companies. Table 5.2 provides a comparison of focal groups, the most detailed level of conceptual tools defined and the third-party auditing schemes available for each CR standard (EU 2009; OECD 2011; Fernández-Muñiz *et al.* 2012; Skaar and Magerholm Fet 2012; GRI and UN 2013; Niskala *et al.* 2013).

In comparison with CR reporting systems, which provide information for sustainability performance measurements and reporting of company business processes, product labels and certificates such as FSC (Forest Stewardship Council) or PEFC (Programme for the Endorsement of Forest Certification) and EPDs (Environmental Product Descriptions) such as the BIFMA (Business and Institutional Furniture Manufacturers Association) standard are used to enhance companies' images and the legitimacy of their operations by confirming that the requirements have been met for the specific sustainability schemes set for their products (e.g. Cobut *et al.* 2013). An important difference between product labels/certificates and EPDs is that product labels and certificates cover assessment of a small number, or just one, of the phases of a product's life-cycle, while EPDs take into account the whole product life-cycle (e.g. Bergman and Taylor 2011). In addition, although the main focus of EPDs is on the ecological aspects of products, social factors (e.g. the health impact of the materials) are also increasingly being included in EPD impact assessments (e.g. Cobut *et al.* 2013).

Table 5.2 Focal groups, level of conceptual tools and third-party auditing schemes for different CR standards

CR standard	*Focal group(s) of organizations*	*Conceptual tools*	*Auditing schemes*
OECD: Guidelines for Multinational Enterprises	Companies aiming, e.g., to ensure that their operations are in harmony with government policies, to strengthen the mutual confidence between enterprises and surrounding societies and to improve the foreign investment climate	Principles[a]	No policies
UN: Global Compact	Companies valuing, embracing, supporting and promoting human rights, labor standards, the environment and anti-corruption	Principles[a]	No policies
AccountAbility: AA1000	Organizations striving to develop an accountable and strategic approach to sustainability with close stakeholder inclusion	Indicators[b]	Voluntary
ISO: 26000	Companies and organizations aiming to translate sustainability principles into effective actions and best practices related to social responsibility	Criteria[c]	No policies
Global Reporting Initiative: GRI Guidelines	Companies and organizations striving to measure, understand and communicate economic, environmental and social information and to use in governance and management	Indicators	Voluntary
EU: EMAS	Public and private organizations attempting to assess, manage and continuously improve their environmental performance	Indicators	Mandatory
ISO: 14001	Organizations aiming to control and reduce their impact on the environment, to implement their environmental policy, to record what has occurred and to learn from experience	Indicators[d]	Voluntary
BSI: OHSAS 18001	Organizations aiming to support and promote good practice in the area of occupational health and safety via systematic and structured management of these issues	Principles[e,f]	Mandatory
SAI: SA8000	Companies aiming to protect and empower human rights and labor laws for all personnel within their sphere of influence including staff employed both by the company as well as by its suppliers, sub-contractors, sub-suppliers, and home workers	Criteria	Mandatory

[a]Principles and indicator-level assessments are in accordance with the GRI Guidelines.
[b]Indicators are to be developed by the reporting organization in dialogue with the stakeholders.
[c]Criteria (called "Clauses" in the measurement system) are complementary with the GRI Guideline and indicator-level measurements are applicable with the GRI Guidelines.
[d]Indicators are mandatory, but they are not specified in the reporting system.
[e]Criteria are compatible with ISO 14001 measurement system.
[f]Principles are quantified into measurable indicators within the auditing scheme.

Stakeholder interaction in the global forest industry

With a focus on global forest industry stakeholders, the following section introduces empirical insights, prevailing CR standards and the normative basis of stakeholder management. The stakeholder groups considered in this analysis include governments, investors, employees, communities, customers, suppliers, consumers and competitors.

According to Vogel (2006), "Civil and government regulation both have a legitimate role to play in improving public welfare. The former reflects the potential for the market for virtues; the latter recognizes its limits." The role of *government* is first and foremost to ensure that the interests of all stakeholder groups are taken into consideration by raising the most critical normative issues to the level of legislation (Donaldson and Preston 1995). A government's role is thus to support – via legislation, tax and funding systems, international standards and labels, voluntary initiatives, capacity building and stakeholder management – the normative base through governmental policies, programs and actions promoting CR among different stakeholders (Albareda *et al.* 2007). Some examples of governmental instruments targeting forest sector CR include the establishment of public procurement policies, setting up governmental certification systems, and financing or co-financing certification activities (e.g. Rametsteiner 2002). The emergence of forest certification schemes in the 1990s is an example of governmental involvement in forest sector CR where private–public partnerships between governments and NGOs played a central role. For example, the FSC formed alliances with governments, NGOs and retailers, while the PEFC endorsed schemes and forged alliances with landowner associations and forest industries. Governments have also played a significant role in the subsequent development of certification standards and the adoption of forest certification schemes (see e.g. McNichol 2002; Overdevest 2010).

Improvements in sustainability can also originate in the private sector before being adopted as public policy, for instance in cases where voluntary industry regulation reinforces and facilitates the development of legislation and serves to "raise the bar" by increasing the level of legal compliance (Vogel 2006). A recent legislative example of industry self-regulation becoming a legal requirement is EU Timber Regulation No 995/2010 (EUTR), which came into effect on 3 March 2013 and prohibits trade in and the import of illegally harvested timber and wood products (European Commission, 2010). The EUTR also recognizes forest and Chain of Custody (CoC) certification and other third-party verification systems for product labelling as valid means of proving the legality and origin of timber, thus promoting the use of voluntary CR reporting tools. Similar initiatives are the Lacey Act in the US and the Illegal Logging Prohibition Bill in Australia.

Traditionally, corporate success has been measured solely in terms of the maximization of value for the *shareholder*, or primary stakeholder, through the creation of wealth; nevertheless, scholars such as Clarkson (1995) have demonstrated that value creation is also dependent on other factors. In the forest industry, some specialized investment banks and socially responsible investors have emerged as new stakeholder groups, establishing demand for CR instruments such as different certification programs (Nikolakis *et al.* 2012). The Carbon Disclosure Project (CDP), which centers on forestry issues, is an example of a non-profit organization providing a sector-independent initiative to enhance CR reporting and stakeholder communication with a specific focus on increasing the commitment of companies to deforestation-free supply chains (e.g. CDP 2014). Similarly, the Dow Jones Sustainability World Index (DJSWI) and NASDAQ OMX CRD Global Sustainability Index are examples of CR reporting initiatives established to provide, information on listed companies' operations for responsible/ethical investors.

The commitment and loyalty of *employees* towards a company can also be seen as a key potential source of competitive advantage by building tacit knowledge and difficult-to-imitate relationships with external stakeholders. The well-being of employees has been addressed by international agreements (the Universal Declaration on Human Rights, the UN Convention on the Rights of the Child, ILO conventions, UN Global Compact principles on human rights), as well as by adopting voluntary occupational health and safety standards (e.g. OHSAS 18000, SA 8000). Information on employees' health and safety, equal rights, wages and other benefits are part of information routinely reported in companies' annual reports, since in many countries this is already part of an employer's statutory duties. More recently, in the forest sector, the social problems rife in factories operating in developing countries, including issues of child labor, poor wages, hazardous working conditions, excessive working hours and discrimination, have given rise to debates on responsible conduct in global supply chains. With the emergence of the concept of extended responsibility in the supply chain, it can be expected that these issues will also continue to be of concern in the future. Nevertheless, these issues are largely untouched, despite the fact that some previous studies on forest sector SMEs have recognized the vital role of employees as a key target audience for CR practices and communication (see Li *et al.* 2014a; Nippala and Lähtinen 2014).

Employees are typically also members of the local *communities* in which forest industry companies operate. In addition, communities consist of the local population, local governments, various labor unions and locally active NGOs. Moreover, communities can define and construct the normative CR base more broadly and independently than governments. Community CR actions involve favorable tax incentives for businesses in the community, disseminating good practices, creating networks, establishing social movements, and awareness campaigns and projects (Albareda *et al.* 2007). The forest sector has a checkered history of conflicts due to either industry expansion into emerging markets and production areas, or disputes regarding the rights of aboriginal people (e.g. Joutsenvirta 2009; Lawrence, 2007). Since the local community plays a vital role in stakeholder management, it would be important to establish CR reporting systems and goals for corporate–community partnerships (see e.g. Esteves and Barclay 2011); however, these have not been generally adopted in the forest industry.

It has become evident in a number of literature reviews that a company's CR performance positively correlates with *customer* behavior towards that company and its products (e.g. Bhattacharya and Sen 2004; Peloza and Shang 2011), implying that sustainability-oriented companies are able to distinguish themselves from competitors, enhance customer satisfaction and loyalty and improve their image. Adoption of CR reporting systems (such as ISO 14001 and EMAS) can be used as examples of an organization's overall strategic commitment to environmental issues. In the European wood products value chain, these systems have mainly been recognized among primary producers, and to some extent among refiners (e.g. Räty *et al.* 2014). Forest-sector-specific standards and certificates include forest and CoC certificates (PEFC, FSC, Real Wood), and primary and refined forest product *suppliers* are especially active in providing forest and CoC certified products. Demand for these CR instruments and products is primarily from industrial customers within the production value chain and from suppliers to the export markets, as well as from wholesalers and retailers, and large-scale pulp and paper product companies (i.e. business-to-business trade), but less from the final *consumers*.

Forest and CoC certification has been criticized for being unable to attract significant price premiums for certified products; however, the fundamental reason has been a lack of consumer demand (e.g. Cai and Aguilar 2013), despite the establishment of on-product eco-labels (e.g. the Nordic eco-label, EU eco-label) or ethical labels (e.g. Fairtrade, country of origin labels) to help end-consumers take environmentally and socially sound purchasing decisions and enhance consumer trust in sustainability claims. More recently, there has been an increase in wood product value chain Environmental Life Cycle Assessment and EPDs (see also the previous section), partly related to the emerging green building initiatives that measure the total environmental impact of products throughout their whole life cycle, from the stand to final-consumer and product disposal.

Finally, the contemporary view of CR in the forest industry has largely lacked consideration of *competitors*, both those inside the industry as well as those in different competing material sectors. Adoption of the Global Reporting Initiative (GRI) and Carbon Disclosure Project can be seen as an attempt to benchmark CR performance in terms of a company's proactive or defensive strategic orientation towards CR (see e.g. Toppinen *et al.* 2012). Nevertheless, although these systems allow comparison of competitors within the sector, most of the reports, analysis and indexes they produce fail to measure performance vis-à-vis competing sectors, such as those material sectors based on oil, concrete and steel.

Company example on prioritizing stakeholder needs

A challenge remains for forest industry companies in defining materiality (common firm–stakeholder interests in CR) and effectively balancing the various needs of different stakeholder groups. An example of firm–stakeholder interests in CR issues is UPM-Kymmene, one of the largest global forest industry companies. In its annual report (UPM 2013), the company describes its stakeholder engagement process in the monitoring of global sustainability megatrends, the mapping of risks and the observation of signs of weakness in its global operating environment. Based on dialogue with its stakeholders, UPM-Kymmene defined four critical stakeholder issues: (1) safety and environmental risk management, (2) sustainable forest management and raw material sourcing, (3) resource efficiency and (4) product characteristics such as safety and eco-labels. Furthermore, these areas became the focus of the company's CR implementation. Recently, UPM-Kymmene radically changed its strategy and re-branded itself as a "Biofore Company," with an increased emphasis on employees and customers as the core internal and external stakeholder groups.

Traditionally, the global discourse on climate change or land management has been less emphasized by the forest industry than by its stakeholders. However, these important stakeholder issues are potential sustainability areas where UPM could improve stakeholder engagement by incorporating them into future company strategies. This could further contribute to value creation and long-term profitability with new stakeholder groups and markets. In contrast, material efficiency has become part of UPM's CR implementation strategy because of the obvious economic benefits to the firm, and probably also due to its being a cost-effective means of seeking stakeholder acceptability.

Success and limitations of CR in the global forest industry

The ultimate success of corporate responsibility naturally varies according to the various CR practices on which a company focuses and the nature of the primary and secondary stakeholder groups and their conflicting needs. Traditionally, in implementing CR in the forest sector, emphasis has been placed on environmental responsibility (e.g. Vidal and Kozak 2008), but social and even cultural (e.g. Lähtinen and Myllyviita 2014) issues have also become more topical, due, in part, to the industry globalization and heightened stakeholder interests mentioned above. In this light, one new feature in the implementation of CR from the European forest industry perspective has been increased interest in building the image of corporations' sustainability via high-profile philanthropic donations unconnected with core (forestry) activities (see e.g. Halme and Laurila 2009). An illustrative example of this is a recent (spring 2014) donation of US$1.3 million by Metsä Group towards the construction of a new children's hospital in Finland, which constitutes the company's single largest philanthropic contribution in its 80-year history.

Nevertheless, as illustrated in the previous two sections, the vast majority of what can be considered forest industry CR comes in the form of applying certificates and standards and guidance from management systems. However, as Rasche (2010) observes, responsibility standards alone can never be a complete solution to the plethora of social and environmental problems experienced today. Indeed, according to de Colle *et al.* (2013), the potential negative effects of CR standards include "obsession with compliance," and the limited ability of standards to innovate and lead systemic change towards sustainability. So is the "standardization drive" too strongly present in the forest industry, and if it is, what does it reveal about awareness of the changing business environment and degree of proactiveness in the sector?

Clearly, in terms of limitations, one must remember that investments to improve corporate sustainability tend to be costly in the short term, while benefits accrue over a longer period of time. Moreover, it is evident that norms and expectations for what constitutes CR for a forestry firm may vary greatly between different contexts and stakeholders. An analysis of trade-offs between competitiveness and sustainability is therefore one of the core issues in pursuing CR (see e.g. van Beurden and Gössling 2008; Orlitzky 2008; Alguinis and Glavas 2012). Although measurement of the economic impact of CR is far from straightforward, the existing reviews and meta-analyses have demonstrated that the relationship between corporate social/environmental performance and financial performance is likely to be positive. However, with the exception of Hansen *et al.* (2013) and Li *et al.* (2014b), research-based evidence on the relationship between CR and economic/financial performance in the forest industry is limited. According to Hansen *et al.* (2013), the economic downturn in the North American forest products industry has even negatively affected company managers' belief in the ability of CR to economically contribute to company performance.

In Europe, contemporary industry perceptions of CR also seem ambivalent: financial hardship has led to the elimination of non-core activities, production downshifting and constant pressure to reduce costs. In this context, the increasing amount of environmental regulation is seen as a threat to the maintenance of current forest industry production levels. For example, in Northern Europe, the EU directive on sulfur emissions in SECA/Baltic Sea marine transportation is a prime example of the perceived negative effect of investment in more sustainable transportation of raw materials and finished goods (e.g. Confederation of European Paper Industries 2013; Gritsenko and Yliskylä-Peuralahti 2013). In conclusion, gaining better understanding of the interplay between sustainability and business performance is an important topic for future research. In addition, ways to improve stakeholder communication and engagement need further consideration, for example by assessing the business opportunities provided by shared firm–stakeholder interests.

Future of CR in the forest industry?

As we all share this great planet, we should also learn to strive towards common goals. Based on our review of the literature and insights into

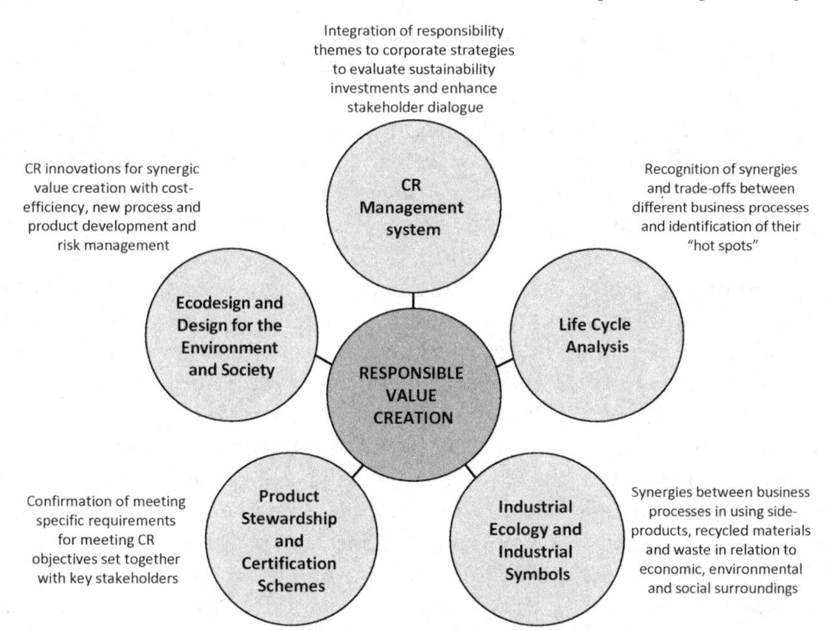

Figure 5.2 Responsible value creation and long-term stakeholder engagement model for the global forest industry (adapted from Sarkis 2012)

industry practices, we believe that sustainable value creation and stakeholder engagement in the forest industry is essentially composed of the following elements (see Figure 5.2): first, the establishment of a coherent CR management system (e.g. decisions on sustainability investments and forms of stakeholder dialogue), and extension of product-level thinking on value creation to include recycling and disposal (e.g. Life Cycle Assessments) and the incorporation of industry byproducts and investments into flexible production systems to produce both inter-firm and inter-sectorial synergies and advance sustainability at a more systemic level; second, aiming at more efficient and transparent communication of the sector's sustainability efforts to its stakeholders, e.g. not only using eco-labels and certificates but also utilizing new innovations for ensuring future competitiveness and the acceptability of operations at a more general level. This can be based on such things as improved energy and resource-efficiency, more synergic value-creation between various stages in the forest-wood value chains and finding solutions to increase customers' quality of life by providing durable and safe products with integrated service components (e.g. wooden houses heated with renewable energy) instead of simply focusing on tangible products (e.g. just wooden houses). Consequently, below we discuss the agenda as we see it from four specific viewpoints: sustainability performance measurement, communication, identified needs of cross-sectorial and stakeholder collaboration, and the need to improve forest industry culture for sustainability.

Naturally, a challenge remains in how to develop *CR performance measurement* systems that can more clearly link implementation of CR practices to traditional corporate objectives (e.g. profitability and growth). One potential concrete business benefit of such systems is the simultaneous recognition of both resource-efficiency gains (e.g. reducing the usage of raw materials, resulting in ecological benefits and decreasing costs) and profit generation potential (e.g. designing business processes to meet specific stakeholder needs). By transparently aligning the implementation of CR practices with the economic objectives of companies, the requirements of heterogeneous stakeholder groups, such as forest owners and consumers, can also be considered in a more balanced way. In addition, questions of equity and the distribution of economic benefits through the whole value chain can also be tackled.

Here more efficient *CR communication* through stakeholder engagement (Clarkson 1995; Manetti 2011) also becomes of increasing importance, as long-term dialogue and development of solid company/stakeholder/local-community relationships are required. For example, the present transition of the pulp and paper industry towards reliance on plantations as sources of fast-growing fiber is typically based on water-intensive, genetically modified monocultures whose long-term effect (see e.g. GRI 2011) on both forest ecosystem services and local-level firm interdependencies with communities is underreported and poorly understood. From the perspective of the communication of sustainability, better understanding of stakeholders and improved communication channels that also target consumers could allow the dissemination of information for benchmarking wood with other material sectors. Eventually, more general awareness and recognition could contribute to improved stakeholder engagement and lead to sustainable leadership being viewed as a competitive advantage in a larger materials market. Here, green governmental procurement policies could also act as a catalyst for promoting CR across sector boundaries.

In line with the model in Figure 5.2, the use of *cross-sectoral collaboration* could be beneficial both for creating innovations and synergies and also for developing the measurement side, in terms of CR indicators and other tools. One of the challenges is how well the traditional wood, pulp and paper driven sector can align itself with the change towards a conservation focused/bio-based economy. Recently, there have been signs in the strategies of some large forest industry companies (see e.g. the recent re-branding of UPM-Kymmene to UPM The Biofore Company) that their future strategic orientation will increasingly expand to include such areas as the production of renewable fuels, chemicals, bio-based fibrils and wood composite materials. From the sustainability perspective, this kind of strategy may bring some benefits, since knowledge in raw-material sourcing and production may be used to develop forest industry compliance with CR requirements, particularly at the beginning of the value chain.

But is there sufficient potential for the strategic commitment of top management and the creation of a *company culture of sustainable leadership* in the forest industry? In line with Heikkurinen and Bonnedahl (2013), we consider that the key question is whether forest industry CR practices mainly deal with a weak form of sustainability. If this is so, there may be a lack of intrinsic motivation to establish long-term stakeholder relationships for implementation of CR, and related corporate strategies would then simply be the result of reactive actions driven by external stakeholder pressures. Forest industry rhetoric is in general very positive towards responsible business conduct (see e.g. WBCSD 2011), but in order to gain more clear value from it, front-runner companies would need to more forcefully distance themselves from the sector's image as an old-fashioned, polluting industry with a traditional mindset. The case of large multinational corporations headquartered mainly in the Nordic countries typifies the dilemma faced by forest companies. This is because, while, according to industry self-perceptions, they are today's front-runners in sustainability (Puumalainen 2013), these very same companies continue to experience negative media coverage of their attempts to develop their activities or expand to emerging low-cost production countries, due to unresolved land ownership issues or the payment of wages below local subsistence levels. Something is thus missing, not only from the viewpoint of industry communication, but also from the perspective of genuine commitment to corporate responsibility.

To meet these challenges – be they related to business diversification, industry conservatism, poor sustainability communication or the underdeveloped image of the sector – what is needed are forward-looking business and sustainability strategies among companies and more efficient and comprehensive communication that targets society as a whole. This would require, however, consideration of the broader institutional settings (e.g. Brammer *et al.* 2012) of forest sector responsibilities and careful evaluation of the rationale and compatibility of short- and long-term business goal-setting. However, since good planets are hard to find, we are optimistic that the forest industry will find both the will and the way to contribute to the future of forests and the well-being of the people of this planet.

References

AccountAbility (2012) 'AccountAbility's AA1000 series', www.accountability.org/standards/index.html, accessed 7 August 2014.

Albareda, L., Lozano, J.M. and Ysa, T. (2007) 'Public policies on corporate social responsibility: the role of governments in Europe', *Journal of Business Ethics*, 74(4), pp. 391–407.

Alguinis, H. and Glavas, A. (2012) 'What we know and don't know about corporate social responsibility: a review and research agenda', *Journal of Management*, 38(4), pp. 932–968.

Barney, J. (1986) 'Organizational culture: can it be a source of sustained competitive advantage?', *Academy of Management Review*, 11(3), pp. 656–665.

Benoît, C., Norris, G.A., Valdivia, S., Ciroth, A., Moberg, A. and Bos, U. (2010) 'The guidelines for social life cycle assessment of products: just in time!', *International Journal of Life Cycle Assessment*, 15(2), pp. 156–163.

Bergman, R. and Taylor, A. (2011) 'EPD – Environmental Product Declarations for wood products – an application of life cycle information about forest products', *Forest Products Journal*, 61(3), pp. 192–201.

Bhattacharya, C.B. and Sen, S. (2004) 'Doing better at doing good: when, why, and how consumers respond to corporate social initiatives', *California Management Review*, 47(1), pp. 9–24.

Bossel, H. (1996) 'Deriving indicators of sustainable development', *Environmental Modeling and Assessment*, 1(4), pp. 193–218.

Brammer, S., Jackson, G. and Matten, D. (2012) 'Corporate social responsibility and institutional theory: new perspectives on private governance', *Socio-Economic Review*, 10(1), pp. 3–28.

Branco, M.C. and Rodrigues, L.L. (2006) 'Corporate social responsibility and resource-based perspectives', *Journal of Business Ethics*, 69(2), pp. 111–132.

BSI (2007) 'Occupational health and safety management systems (OHSAS 18001:2007)', www.bsigroup.com/en-GB/ohsas-18001-occupational-health-and-safety, accessed 7 August 2014.

Cai, Z. and Aguilar, F. X. (2013) 'Meta-analysis of consumer's willingness-to-pay premiums for certified wood products', *Journal of Forest Economics*, 19(1), pp. 15–31.

CDP (2014) 'Deforestation-free supply chains: From commitments to action', www.cdp.net/CDPResults/CDP-global-forests-report-2014.pdf, accessed 18 November 2014.

Clarkson, M.B.E. (1995) 'A stakeholder framework for analyzing and evaluating corporate social performance', *Academy of Management Review*, 20(1), pp. 92–117.

Cobut, A., Beauregard, R. and Blanchet, P. (2013) 'Using life cycle thinking to analyze environmental labeling: the case of appearance wood products', *International Journal of Life Cycle Assessment*, 18(3), pp. 722–742.

Confederation of European Paper Industries (2013) 'Sustainability report 2013', www.cepi-sustainability.eu/uploads/Full_sustainability2013.pdf, accessed 7 August 2014.

Crane, A., Matten, J. and Spence, L. (2013) 'Corporate social responsibility: in a global context', in A. Crane, D. Matten and L.J. Spence (eds) *Corporate Social Responsibility: Readings and Cases in a Global Context*, Abingdon: Routledge.

Dahlsrud, A. (2008) 'How corporate social responsibility is defined: an analysis of 37 definitions', *Corporate Social Responsibility and Environmental Management*, 15(1), pp. 1–13.

de Colle, S., Henriques, A. and Sarasvathy, S. (2013) 'The paradox of corporate social responsibility standards', *Journal of Business Ethics*, DOI 10.1007/s10551-10013-1912-y.

Donaldson, T. and Preston, L.E. (1995) 'The stakeholder theory of the corporation – concepts, evidence, and implications', *Academy of Management Review*, 20(1), pp. 65–91.

Eccles, R., Ioannou, I. and Serafeim, G. (2011) 'The impact of corporate culture of sustainability on corporate behavior and performance', Harvard Business School Working Paper, 12–035.

Esteves, A.M. and Barclay, M. (2011) 'New approaches to evaluating the performance of corporate–community partnerships: a case study from the minerals sector', *Journal of Business Ethics*, 103(2), pp. 189–202.

EU (2009) 'Regulation of the European Parliament and of the Council on the voluntary participation by organisations in a Community eco-management and audit scheme (EMAS)', http://eur-lex.europa.eu/legal-content/EN/TXT/PDF/?uri=CELEX:32009R1221&from=EN, accessed 8 August 2014.

European Commission (2001) 'Green Paper "Promoting a European framework for Corporate Social Responsibility"', csr-in-commerce.eu/data/files/resources/717/com_2001_0366_en.pdf, accessed 8 August 2014.

European Commission (2010) 'Regulation of the European Parliament and of the Council laying down the obligations of operators who place timber and timber products on the market (EU No. 995/2010)', http://ec.europa.eu/environment/forests/timber_regulation.htm, accessed 8 August 2014.

European Commission (2011) 'A renewed EU strategy 2011–2014 for Corporate Social Responsibility', http://eur-lex.europa.eu/LexUriServ/LexUriServ.do?uri=COM:2011:0681:FIN:EN:PDF, accessed 8 August 2014.

Fernández-Muñiz, B., Montes-Peón, J.M. and Vázquez-Ordás, C.J. (2012) 'Safety climate in OHSAS 18001-certified organisations: antecedents and consequences of safety behaviour', *Accident Analysis and Prevention*, 45, pp. 745–758.

Finnveden, G., Hauschild, M.Z., Ekvall, T., Guinee, J., Heijungs, R. and Hellweg, S. (2009) 'Recent developments in life cycle assessment', *Journal of Environmental Management*, 91(1), pp. 1–21.

Freeman, R. (1984) *Strategic Management: A Stakeholder Approach*, Marshfield, MA: Pitman.

Galbreath, K. (2009) 'Building corporate social responsibility into strategy', *European Business Review*, 21(2), pp. 109–127.

Gonzaléz-Garcia, S., Moreira, M.T., Dias, A.C. and Mola-Yudego, B. (2014) 'Cradle-to-gate life-cycle assessment of forest operations in Europe: environmental and energy profiles', *Journal of Cleaner Production*, 66(1), pp. 188–198.

GRI (2011) 'Global Reporting Initiative. Sustainability reporting guidelines', www.globalreporting.org/resourcelibrary/G3.1-Guidelines-Incl-Technical-Protocol.pdf, accessed 8 August 2014.

GRI and ISO (2014) 'GRI G4 Guidelines and ISO 26000:2010. How to use the GRI G4 Guidelines and ISO 26000 in conjunction', www.iso.org/iso/iso-gri-26000_2014-01-28.pdf, accessed 8 August 2014.

GRI and UN (2013) 'Making the Connection: Using the GRI G4 Guidelines to communicate progress on the UN Global Compact principles', www.globalreporting.org/resourcelibrary/UNGC-G4-linkage-publication.pdf, accessed 8 August 2014.

Gritsenko, D. and Yliskylä-Peuralahti, J. (2013) 'Governing shipping externalities: Baltic ports in the process of SOx emission reduction', *Maritime Studies*, 12(1), pp. 1–21.

Halme, M. and Laurila, J. (2009) 'Philanthropy, integration or innovation? Exploring the financial and societal outcomes of different types of corporate responsibility', *Journal of Business Ethics*, 84(3), pp. 325–339.

Hansen, E., Nybakk, E. and Panwar, R. (2013) 'Firm performance, business environment, and outlook for social and environmental responsibility during the economic downturn: findings and implications from the forest sector', *Canadian Journal of Forest Research*, 43(12), pp. 1137–1144.

Hart, S. (1995) 'A natural-resource-based view of the firm', *Academy of Management Review*, 20(4), pp. 986–1014.

Heikkurinen, P. and Bonnedahl, K. (2013) 'Corporate responsibility for sustainable development: a review and conceptual comparison of market- and stakeholder-oriented strategies', *Journal of Cleaner Production*, 43, pp. 191–198.

Husgafvel, R., Watkins, G., Linkosalmi, L. and Dahl, O. (2013) 'Review of sustainability management initiatives within Finnish forest products industry companies – translating EU level steering into proactive initiatives', *Resources, Conservation and Recycling*, 76, pp. 1–11.

ISO (2004) 'ISO 14000 – Environmental management'. International Organization for Standardization, www.iso.org/iso/home/standards/management-standards/iso14000.htm, accessed 8 August 2014.

ISO (2010) 'ISO 26000 – Social responsibility'. International Organization for Standardization, www.iso.org/iso/home/standards/iso26000.htm, accessed 8 August 2014.

Joutsenvirta, M. (2009) 'A language perspective to environmental management and corporate responsibility'. *Business Strategy Environment* , 18, 240–253.

Korhonen, J. and Snäkin, J. (2005) 'Analysing the evolution of industrial ecosystems: concepts and application', *Ecological Economics*, 52(2), pp. 169–186.

Kozak, R. (2013) 'What now Mr. Jones? Some thoughts about today's forest sector and tomorrow's great leap forward', in R. Panwar, E. Hansen and R. Vlosky (eds) *The Global Forest Sector: Changes, Practices and Prospects*. Boca Raton, FL: CRC Press,. pp. 431–445.

Kurucz, E., Colbert, B. and Wheeler, D. (2008) 'The business case for corporate social responsibility', in A. Crane, A. McWilliams, J. Moon and D. Siegel (eds) *The Oxford Handbook of Corporate Social Responsibility*. Oxford: Oxford University Press.

Lähtinen, K. and Myllyviita, T. (2014) 'Cultural sustainability in reference to the Global Reporting Initiative (GRI) guidelines – case forest bioenergy production in North Karelia, Finland', *Journal of Cultural Heritage Management and Sustainable Development*. In press.

Lähtinen, K. and Toppinen, A. (2008) 'Financial performance in Finnish large- and medium-sized sawmills: the effects of value-added creation and cost-efficiency seeking', *Journal of Forest Economics*, 14(4), pp. 289–305.

Lähtinen, K., Myllyviita, T., Leskinen, P. and Pitkänen, S. (2014) 'A systematic literature review on indicators to assess local sustainability of forest energy production', *Renewable and Sustainable Energy Reviews*, 40, pp. 1202–1216.

Lähtinen, K., Samaniego Vivanco, D. and Toppinen, A. (2014) 'Designers' wooden furniture ecodesign implementation in Scandinavian country-of-origin (COO) branding', *Journal of Product & Brand Management*, 23(3), pp. 180–191.

Lawrence, R. (2007) 'Corporate social responsibility, supply-chains and Saami claims: tracing the political in the Finnish forestry industry', *Geographical Research*, 45(2), pp. 167–176.

Li, N., Toppinen, A. and Lantta, M. (2014a) 'Corporate responsibility in the forest industry SMEs: evidence from China and Finland', online first version in *Journal of Small Business Management*, DOI: 10.1111/jsbm.12136.

Li, N., Puumalainen, K. and Toppinen, A. (2014b) 'Managerial perceptions of corporate social and financial performance in global forest industry', *International Forestry Review*, 16(3), pp. 319–338.

Litz, R.A. (1996) 'A resource-based-view of the socially responsible firm: stakeholder interdependence, ethical awareness, and issue responsiveness as strategic assets', *Journal of Business Ethics*, 15(12), pp. 1355–1363.

Lozano, R. and Huisingh, D. (2011) 'Inter-linking issues and dimensions in sustainability reporting', *Journal of Cleaner Production*, 19(2–3), pp. 99–107.

Macombe, C., Leskinen, P., Feschet, P. and Antikainen, R. (2013) 'Social life cycle assessment of biodiesel production at three levels: a literature review and development needs', *Journal of Cleaner Production*, 52, pp. 205–216.

Malik, M. (2014) 'Value-enhancing capabilities of CSR: a brief review of contemporary literature', *Journal of Business Ethics*, 127(2), 419–438.

Manetti, G. (2011) 'The quality of stakeholder engagement in sustainability reporting: empirical evidence and critical points', *Corporate Social Responsibility and Environmental Management*, 18(2), pp. 110–122.

Mayer, A.L., Kauppi, P.E., Angelstam, P.K., Zhang, Y. and Tikka, P.M. (2005) 'Importing timber, exporting ecological impact', *Science*, 308(5720), pp. 359–360.

McNichol, J.H. (2002) 'Contesting governance in the global marketplace: a sociological assessment of business–NGO partnerships to build markets for certified wood', PhD thesis, University of California at Berkeley.

Mendoza, G.A. and Prabhu, R. (2000) 'Development of a methodology for selecting criteria and indicators of sustainable forest management: a case study on participatory assessment', *Environmental Management*, 26(6), pp. 659–673.

Mikkilä, M. and Toppinen, A. (2008) 'Corporate responsibility reporting by large pulp and paper companies', *Forest Policy and Economics*, 10(7–8), pp. 500–506.

Myllyviita, T., Holma, A., Antikainen, R., Lähtinen, K. and Leskinen, P. (2012) 'Assessing environmental impacts of biomass production chains – application of life cycle assessment (LCA) and multi-criteria decision analysis (MCDA)', *Journal of Cleaner Production*, 29–30, pp. 238–245.

Myllyviita, T., Leskinen, P., Lähtinen, K., Pasanen, K., Sironen, S., Kähkönen, T. and Sikanen, L. (2013) 'Sustainability assessment of wood-based bioenergy – a methodological framework and a case-study', *Biomass and Bioenergy*, 59, pp. 293–299.

Nikolakis, W., Cohen, D.H. and Nelson, H.W. (2012) 'What matters for socially responsible investment (SRI) in the natural resources sectors? SRI mutual funds and forestry in North America', *Journal of Sustainable Finance and Investment*, 2(2), pp. 136–151.

Nippala, J. and Lähtinen, K. (2014) 'Corporate social responsibility and sustainability in North Carolina's small and medium-sized forest products companies', Gronen Research Conference 2014, Helsinki.

Niskala, M., Pajunen, T. and Tarna-Mani, K. (2013) 'Yritysvastuu, raportointi ja laskentaperiaatteet', (in Finnish), KHT Pro, Helsinki.

OECD (2011) 'OECD Guidelines for Multinational Enterprises', www.oecd.org/daf/inv/mne/48004323.pdf, accessed 8 August 2014.

Orlitzky, M. (2008) 'Corporate social performance and financial performance: a research synthesis', in A. Crane, A. McWilliams, D. Matten, J. Moon and D. Siegel (eds) *Oxford Handbook of Corporate Social Responsibility*. Oxford: Oxford University Press.

Overdevest, C. (2010) 'Comparing forest certification schemes: the case of ratcheting standards in the forest sector', *Socio-Economic Review*, 8(1), pp. 47–76.

Panwar, R. and Hansen, E. (2007) 'The standardization puzzle: an issue management approach to understand corporate responsibility standards for the forest products industry', *Forest Products Journal*, 57(12), pp. 86–91.

Peloza, J. and Shang, J. (2011) 'How can corporate social responsibility activities create value for stakeholders? A systematic review', *Journal of the Academy of Marketing Science*, 39(1), pp. 117–135.

Porter, M.E. and Kramer, M.R. (2006) 'Strategy and society', *Harvard Business Review*, 84(12), pp. 78–92.

Porter, M.E. and Kramer, M.R. (2011) 'Creating shared value', *Harvard Business Review*, 89(1–2), pp. 62–77.

Puumalainen, K. (2013) 'Yhteiskuntavastuun merkitys yritykselle' (in Finnish). www.metsateollisuus.fi/mediabank/183.pdf, accessed 8 August 2014.

Rametsteiner, E. (2002) 'The role of governments in forest certification – a normative analysis based on new institutional economics theories', *Forest Policy and Economics*, 4(3), pp. 163–173.

Rasche, A. (2010) 'The limits of corporate responsibility standards', *Business Ethics: A European Review*, 19(3), pp. 280–291.

Rasche, A. and Esser, D.E. (2006) 'From stakeholder management to stakeholder accountability – applying Habermasian discourse ethics to accountability research', *Journal of Business Ethics*, 65(3), pp. 251–267.

Räty, T., Toppinen, A., Roos, A., Riala, M. and Nyrud, A.Q. (2014) 'Environmental policy in the Nordic wood product industry: insights into firms' strategies and communication', *Business Strategy and the Environment*, DOI: 10.1002/bse.1853.

Robèrt, K.H., Schmidt-Bleek, B., de Larderel, J.A., Basile, G., Jansen, J.L. and Kuehr, R. (2002) 'Strategic sustainable development – selection, design and synergies of applied tools', *Journal of Cleaner Production*, 10(3), pp. 197–214.

SAI (2008) 'Social Accountability 8000 Standard', www.sa-intl.org/_data/n_0001/resources/live/2008StdEnglishFinal.pdf, accessed 8 August 2014.

Sarkis, J. (2012) 'A boundaries and flows perspective of green supply chain management', *Supply Chain Management: An International Journal*, 17(2), pp. 202–216.

Skaar, C. and Magerholm Fet, A. (2012) 'Accountability in the value chain: from environmental product declaration (EPD) to CSR product declaration', *Corporate Social Responsibility and Environmental Management*, 19(4), pp. 228–239.

Toppinen, A., Li, N., Tuppura, A. and Xiong, Y. (2012) 'Corporate responsibility and strategic groups in the forest-based industry: exploratory analysis based on the Global Reporting Initiative (GRI) framework', *Corporate Social Responsibility and Environmental Management*, 19(4), pp. 191–205.

UPM (2013) 'Annual Report', www.upm.com/EN/INVESTORS/Documents/UPM_Annual_Report_2013.pdf, accessed 8 August 2014.

van Beurden, P. and Gössling, T. (2008) 'The worth of values – a literature review on the relation between corporate social and financial performance', *Journal of Business Ethics*, 82(2), pp. 407–424.

Vidal, N.G. and Kozak, R.A. (2008) 'Corporate responsibility practices in the forestry sector: definitions and the role of context', *The Journal of Corporate Citizenship*, 31, pp. 49–59.

Vogel, D. (2006) *The Market for Virtue: The Potential and Limits of CSR*, Washington DC: Brooking Institution Press.

von Geibler, J., Kristof, K. and Bienge, K. (2010) 'Sustainability assessment of entire forest value chains: integrating stakeholder perspectives and indicators in decision support tools', *Ecological Modelling*, 221(18), pp. 2206–2214.

WBCSD (2011) 'A Guide to Corporate Ecosystem Valuation. A framework for improving corporate decision making', www.wbcsd.org/pages/edocument/edocumentdetails.aspx?id=104&nosearchcontextkey=true, accessed 8 August 2014.

6 Opportunities and challenges in community forest tenure reform

Alexandre Corriveau-Bourque, Jenny Springer, Andy White and D. Bryson Ogden

Introduction

Few issues are as political as the rights to the world's remaining forest lands. Forests are viewed by a wide range of actors as a source of timber, fiber, food, fuel, medicine, carbon storage, biodiversity, spirituality, and as sites of cultural belonging. Vast mineral, gas, and oil resources are also found beneath the world's forests. As populations and incomes increase around the world, pressures are growing on this shrinking, yet increasingly important, forest estate and the resources it contains. To understand the current contestation for these resources, it is important to understand the following questions: Who "owns" or "controls" forest resources? How is this contestation manifesting itself? And what are some of the implications of this contestation for local peoples, governments, investors, and the forests themselves?

Over the past decades, Indigenous Peoples and other forest communities have substantially increased the proportion of the world's forests they legally own or control, and are mobilizing in new and more effective ways to assert and defend their rights. While many lands and resource rights claimed by these peoples remain unrecognized or are circumvented, pressure is mounting on policy makers. Indigenous Peoples and local communities have been at the forefront of developing and using a number of tools to advance and defend their tenure rights over the past decade, including: the United Nations Declaration on the Rights of Indigenous People (UNDRIP); the mainstreaming of Free, Prior, and Informed Consent (FPIC); the development and wide adoption of the Voluntary Guidelines on the Responsible Governance of Tenure (VGGTs); the legal reforms catalyzed by processes such as REDD+ and the negotiation of Voluntary Partnership Agreements (VPA) between governments and the European Union to ensure compliance with Forest Law Enforcement, Governance, and Trade (FLEGT) regulations; and the opening of fora for the mediation of communities' grievances against company abuses in the Forest Sustainability Council (FSC) and Roundtable for Sustainable Palm Oil (RSPO). National and international judicial systems have also increasingly become vehicles for Indigenous Peoples and local communities to challenge abuses and to establish legal benchmarks for recognizing their rights.

With these increased legal rights and other tools, Indigenous Peoples and local communities have been able to leverage unprecedented power and gain access to policy conversations about the future use, management, and preservation of resources that had formerly excluded them. As these developments gain momentum, the costs for governments and private sector investors of resisting or ignoring the groundswell of indigenous and community mobilization are rising. Furthermore, there is a growing base of evidence that demonstrates that local rights to land and forest resources are essential for the meaningful achievement of economic development, conservation, and climate change mitigation goals (Chhatre and Agarwal, 2009; Larson *et al.*, 2010; Porter-Bolland *et al.*, 2012; Nolte *et al.*, 2013).

The purpose of this chapter is to present recent data on trends and patterns in global forest tenure, based primarily on research conducted by the Rights and Resources Initiative, and to establish the importance for governments, investors, and conservation actors to respect and strengthen community rights. Understanding these forest tenure trends and needs will allow us to better understand progress in the recognition of indigenous and community forest rights, as well as gaps and the further actions needed to ensure an equitable and sustainable future for forests and the people who depend on them.

Forest tenure typology

The 2002 report, *Who Owns the World's Forests?* (White and Martin, 2002), presented a typology of four categories of statutory forest tenure rights and established a baseline for assessing changes in the extent of forest area in each of these categories over time. The report found that while governments retained legal ownership over more than three-quarters of the global forest estate, governments had been increasingly recognizing communities' rights to forest land since the 1980s. Sunderlin *et al.* (2008) updated this analysis and found a continued transition from state ownership to forest ownership or control by Indigenous Peoples and local communities.

The typology used for this analysis of changes in forest tenure from 2002 to 2013 is based on the one used by White and Martin (2002) and Sunderlin *et al.* (2008), but has been slightly updated to reflect new data and interpretations. The tenure categories for the analysis presented in this chapter are defined as follows:

- *Forest land administered by governments:* This category includes all forest land that is legally claimed as exclusively belonging to the state. It includes areas where community rights are limited to basic access or withdrawal rights that can be legally extinguished with relative ease by the state.

- *Forest land designated by governments for Indigenous Peoples and local communities:* Some rights to forests under this category have been recognized by governments on a conditional basis for Indigenous Peoples and local communities. While rights-holders have some level of "control" exercised through management and/or exclusion rights over forests, they lack the full legal means to ensure the security of their claims to forests (i.e. having all three rights to exclude, to due process and compensation, and to retain rights for an unlimited duration).[1]
- *Forest land owned by Indigenous Peoples and local communities:* Forests are considered to be "owned" where communities' legal rights are unlimited in duration, they have the legal right to exclude outsiders from using their resources, and they are entitled to due process and compensation in the face of extinguishment by the state of some or all of their rights. In this analysis, alienation rights are not considered to be essential for community ownership.
- *Forest land owned by individuals and firms:* In these areas, individuals and firms have full legal rights of ownership of forest land.

Rights to forest land acquired by individuals and corporations through long-term leases or concessions are not captured by this typology because individuals and firms may acquire time-bound rights to the forests under any one of these four categories. The true extent of these leases is not yet fully understood on a global scale, nor is the extent of the overlap of these leases with different forms of community rights or individual and firm ownership. The consequences of these overlaps on community claimed, controlled, and owned land will be discussed later in this chapter.

Global forest tenure transition: 2002–2013

The data below present the aggregate findings on changes in the area of forest land from 2002 to 2013 in 40 countries,[2] 33 of which are low- and middle-income countries (LMICs).[3] The 40 countries represent 82% of the global forest area and the 33 countries represent 85% of the forest area in LMICs.

Figure 6.1 demonstrates that governments still overwhelmingly claim control over forest land at the global scale. In four of the eight[4] most-forested countries (by area), governments retain legal administrative control and ownership over at least 90% of their respective forest estates. The Russian Federation alone encompasses nearly 20% of the global forest estate and, by law, all of its forests remain "administered by government." The Democratic Republic of the Congo also has 100% of its forests under government administration. Indonesia and Canada retain 96% and nearly 92% of their

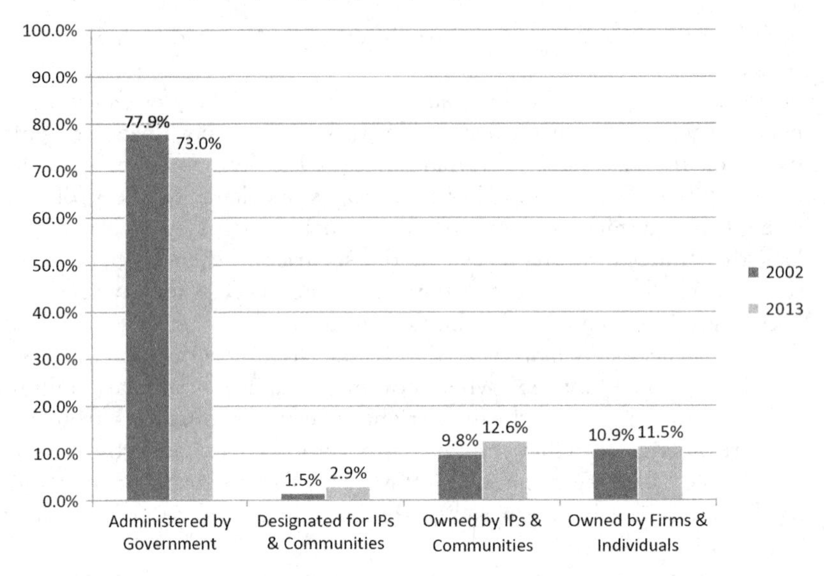

Figure 6.1 Global change in statutory forest land tenure, 2002–2013 (%)
Source: RRI (2014)

respective forests under government control. Together, these four countries contain over a third of the world's forests and nearly 57% of the area under government administration. This means that the absence of large-scale tenure reforms in these countries presents major impediments to global progress in the recognition of local rights to forest land.

Even in the absence of major shifts in four countries with sizable forest areas, the total forest area under the legal ownership or control of Indigenous Peoples (IPs) and local communities increased from 383 Mha (just over 11% of global forest area) in 2002 to over 511 Mha (15.5%) in 2013. Over the same period, the proportion of the forests owned by individuals and firms increased by less than 1%.

By 2013, 31 countries out of the 40 had some form of government recognition of community rights to forests. Of these 31 countries, 27 recorded an increase in the forest area under legal community ownership or control. Nine of these countries[5] had not implemented any form of recognition of community rights to forest lands in 2002, meaning that some reforms were implemented for the first time during this period. These reforms were overwhelmingly in the "designated for Indigenous Peoples and local communities" category, with significant limitations on the security of these rights.

Figure 6.2 shows that during the period from 2002 to 2013 most of the global forest tenure transition towards legal community control and ownership took place in LMICs. In fact, almost all (97%) of the global change in the recognition of community rights over the 2002–2013 period took place in LMICs, with the bulk in Latin America.

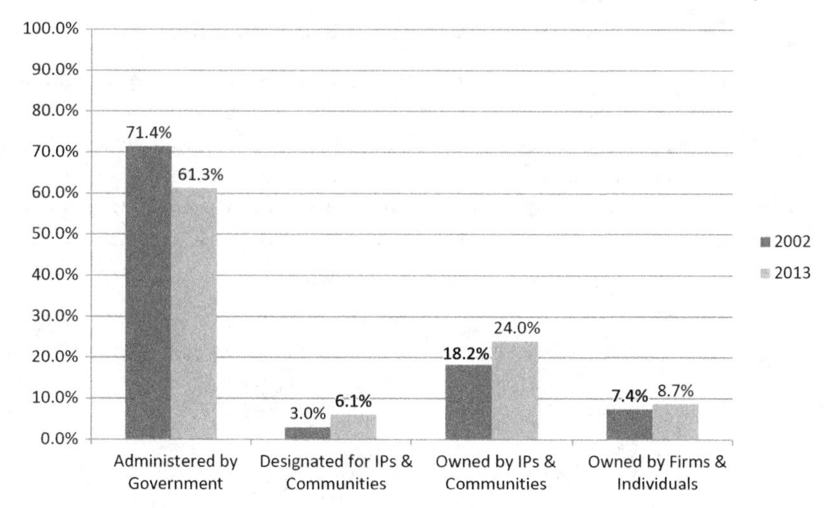

Figure 6.2 Change in statutory forest land tenure in LMICs, 2002–2013, in millions of hectares
Source: RRI (2014)

More specifically, the total forest area under legal community ownership or control in LMICs rose from just over 353 Mha (just over 21% of forest area) in 2002 to at least 478 Mha (30%) in 2013. This equates to an increase of at least 125 Mha of forests in which communities' rights have been recognized. More than 62% of these 125 Mha are owned by communities.

Key countries in the forest tenure transition

From a global perspective, the forest tenure transition is clearly progressing, particularly in LMICs. However, five countries – Bolivia, Brazil, China, Colombia, and Peru – account for the majority of the increase in forest area under indigenous and local community ownership recorded between 2002 and 2013. While not highly visible in global aggregates, some countries with smaller forest areas, such as the Philippines and Honduras have also substantially increased the proportion of their forest land owned by Indigenous Peoples and local communities since 2002.

Of the total forest area legally owned by Indigenous Peoples and local communities in 2013, 80% is found in only five countries. China and Brazil alone account for 55% of the global area, while Colombia, Mexico, and Papua New Guinea account for another 25%.

Of the forests designated for Indigenous Peoples and other communities in 2013, 84% are found in Brazil, India, and Tanzania, with the bulk of the increase in this tenure category in the period 2002–2013 taking place in Brazil and India. Smaller countries, such as Guyana, Nepal, and the Gambia, have also substantially increased the proportion of their forest land designated for Indigenous Peoples and local communities since 2002.

A regional comparison of the forest tenure transition

In light of the more substantial progress in LMICs, and the importance of these countries for global development goals, a regional comparison of trends across sub-Saharan Africa, Asia, and Latin America was also conducted. This analysis reveals considerable regional variation in statutory recognition of forest land rights. As Figure 6.3 shows, the implementation of reforms in sub-Saharan Africa is lagging far behind those in Latin America

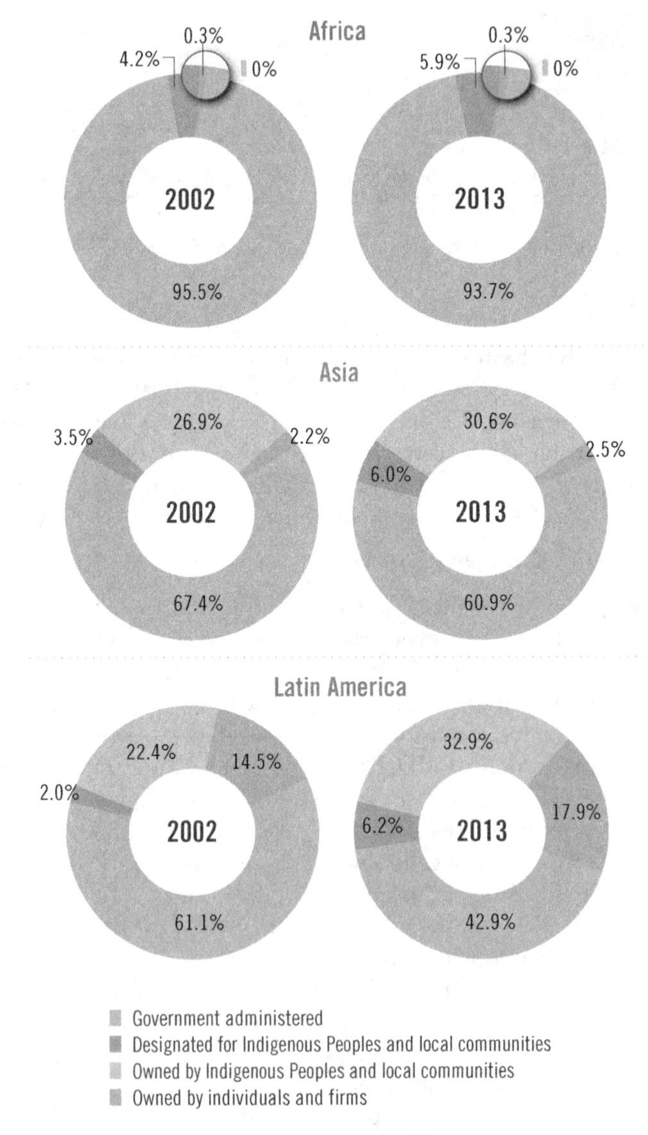

Figure 6.3 Statutory recognition of forest tenure, by region

and Asia; though closer inspection within the regions further nuances this observation.

Forest tenure transition in sub-Saharan Africa

Of the 12 countries with complete data in this region, five failed to recognize communities' rights to forest lands.[6] Of the remaining seven countries where community rights have been recognized, tenure reforms have affected less than 6% of forest area in most countries. Only Tanzania and the Gambia exceeded this proportion.

Overall, as of 2013, less than 6% of forests in sub-Saharan Africa were "designated for Indigenous Peoples and local communities" (see Figure 6.3). The implementation of Tanzania's Village Land Act (1999) and Forest Act (2002) account for over 89% of this area.

Furthermore, there is no recorded area under community "ownership" in Africa. This partly reflects a lack of data for the two countries – Mozambique and Liberia – that have enacted statutory frameworks recognizing community ownership of forest land,[7] but could not be included within the set of countries due to the absence of reliable data.[8] The forest area owned by communities in these countries may be substantial because these laws recognize the rights of communities regardless of whether or not formal titles exist; however, the extent of this area is not yet known.

Nevertheless, even if the entire forest estate of these two countries is recognized under community ownership, there would still be very limited recognition of community rights in the Africa region, due to limited implementation of legal reforms in the Congo Basin,[9] where states retain legal administrative control over 99% of the region's forest estate. Nearly 68% of the forests in sub-Saharan Africa are in the Congo Basin.

However, the lack of legal recognition in any of these countries does not indicate that these lands are uncontested; quite the contrary, most are claimed and directly managed under customary tenure systems (Alden Wily, 2012). The weak recognition and protection of communities' rights by governments is a profound source of economic and political insecurity for these communities and contributes heavily to their political marginalization and poverty, in addition to environmental degradation (Okereke and Dooley, 2010).

While communities in these countries may not have recognized legal rights to control or own their lands, many hold rights through international norms, such as ILO 169,[10] the UNDRIP, and the VGGTs, or through national laws to consultation and/or compensation. However, even such marginal rights are routinely ignored, with devastating consequences for communities, and with potential impacts on governments and private sector investors as well, as explored below.

Forest tenure transition in Asia

Of the 12 countries with complete data in the Asia region, three countries recorded increases in the area owned by communities while nine recorded increases in the area recognized as "designated for Indigenous Peoples and local communities" between 2002 and 2013. By 2013, all 12 countries had implemented some form of community tenure regime; however, this implementation has affected less than 4% of the countries' forests in seven of these countries.

As of 2013, nearly 31% of the forests in Asia were under the ownership of Indigenous Peoples and local communities, and 6% were under community control (see Figure 6.3). However, 78% of the forests owned by Indigenous Peoples and local communities in Asia are found in China[11] within rural collectives. If China is excluded from the set of countries, only 10% of the region's forest land is under community ownership. Similarly, India represents nearly 82% of the regional share of forest land "designated for Indigenous Peoples and local communities." At the same time, the size of China and India should not overshadow the extent of recognition in smaller countries such as Nepal, the Philippines, and Papua New Guinea, which have implemented recognized community rights to 32%, 39%, and 97% of their respective forest areas. Even though the recognition of rights in Papua New Guinea has been extensive, these reforms have not been complemented by improved governance, and therefore communities' rights in the country remain under threat (Oakland Institute, 2013).

Only a third of the countries in the Asia set of countries have implemented tenure reforms recognizing community ownership of forest land while 10 out of the 12 countries have implemented tenure regimes recognizing more limited degrees of community control. In this latter set of countries, seven have implemented reforms recognizing community rights in 3% or less of the total forest area.

This absence of recognition of Indigenous Peoples and local community rights is particularly stark in peninsular Southeast Asia, where states retain legal control over 99% of forest land. In archipelagic Southeast Asia, governments retain control over at least 73% of forest land, which is on a par with the global recognition. However, the progress in the Philippines and Papua New Guinea obscures the lack of recognition of community rights in Indonesia, which remains at about 1% of the total forest area. Reliable data were not available for Malaysia. Large-scale forest tenure reforms in the countries of these two sub-regions would therefore be needed to shift the balance of government and community forest rights in Asia.

As in sub-Saharan Africa, the large areas of forests under government administration in Asia are deeply contested. For instance, in Indonesia, of approximately one million hectares formally recognized to be under community control in Indonesia, most are "designated for communities" in the form of "village plantations." However, the Indigenous Peoples' Alliance of the Archipelago (AMAN), challenges this figure, estimating that, nationally, there are approximately 40 Mha of customary land in Adat villages with contiguous (forest and non-forest) natural resource areas.[12]

Forest tenure transition in Latin America

Many Latin American countries have implemented large-scale forest tenure reforms recognizing the rights of Indigenous Peoples and local communities, and tenure reforms have been more widely distributed across countries than in other regions. In the period 2002–2013, eight of the nine country cases recorded increases in the area recognized under community rights, accounting for an 85 Mha total increase in the area under statutory community control or ownership. This represents nearly 66% of the global increase in area under community ownership or control from 2002 to 2013.

In Latin America, communities now own nearly 33% of forests and control more than 6% of all forests (see Figure 6.3). In the seven countries that recognize community ownership, community-owned forests range from just over 10% to nearly 70% of their countries' respective forest areas. Of the two countries in the set that only recognize community control, Guyana's reforms recognize community control of nearly 17% of forests, while Suriname's reforms cover less than 4%.

However, even within areas formally titled to indigenous communities, there is often contestation with other land uses – especially when these community lands overlap with state-designated protected areas and oil and gas concessions. A study on extractive industries in Guatemala, Colombia, Panama, and Peru found that investments in all four countries contributed to community displacement, large-scale deforestation and the destruction of local livelihoods (Flórez, 2013), driven by the government retaining legal control over subsoil resources.

To put this tension into context, as of 2011, approximately 48 Mha of oil and gas concessions had been granted in the Peruvian Amazon, covering 61% of the forest. These concessions have been found to overlap with four territorial reserves, five communal reserves, and at least 70% of native communities – many of which reject the presence of mining and oil companies (Espinoza and Feather, 2011). Furthermore, in January 2014, the Peruvian government announced the expansion of the Camisea gas project in the Peruvian Amazon, and three-quarters of the area under this concession is inside a territory that was established by the government for Indigenous Peoples living in voluntary isolation (REDD-Monitor, 2014).

A slowdown in the recognition of community rights

A comparison of the pace of recognition of community forest rights in LMICs from 2002 to 2008, and from 2008 to 2013, reveals a slowdown in the recognition of rights. The slowdown was particularly noticeable in the indigenous and local community "ownership" category, where the recorded increase from 2002 to 2008 was 66.8 Mha and the recorded increase from 2008 to 2013 was only 11.2 Mha. In the "designated for Indigenous Peoples and local communities" category, the recorded increase from 2002 to 2008 was 26.8 Mha, while from 2008 to 2013 the recorded increase was 19.7 Mha.

In part, this slowdown may be attributed to the consolidation of reform processes among the large forested countries of Latin America, where most of the implementation took place. Many of these countries began their reform processes in the 1980s and 1990s. Certainly, even in these countries, significant claims remain to be recognized. For instance, in Brazil, only 207 Afro-Brazilian (Quilombo) communities have been issued titles to their lands since 1988, while more than 1,200 claims are pending (Carniero, 2014). In Peru, the Indigenous Peoples' organization, Asociacion Interetnica de Desarrollo de la Selva Peruana (AIDESEP), estimates that at least 20 Mha[13] of additional land, much of it in the Peruvian Amazon, is claimed by Indigenous Peoples but remains unrecognized by the government (Espinoza and Feather, 2011, 8).

Within the overall trend of a slowdown, some lower- and middle-income countries have continued to record greater increases in 2008–2013 than in 2002–2008. These include Honduras, India, and the Philippines in the ownership category, and Guyana, Gabon, Honduras, Lao PDR, Nepal, Tanzania, and Vietnam in the designated for use category. While the increase in some of these countries was incremental, the changes offer hope for potential to push for greater recognition, to strengthen the rights already recognized, and to increase the ability of communities to exercise those rights in law and practice.

Countervailing forces – What future for the recognition of rights?

The slowdown in the recognition of Indigenous Peoples' and communities' rights to forest from 2008 to 2013 coincides with a range of converging global processes and trends. On the one hand, 2007 marked the adoption of UNDRIP by the United Nations General Assembly, as well as the adoption of several major decisions on REDD+ during the 13th Conference of the Parties to the United Nations Framework Convention on Climate Change (COP 13), which was widely seen as the launching point of REDD+. These developments were commonly expected to serve as catalysts for the broader recognition of local rights. In the context of REDD+, 27 of the 35 national REDD+ programs acknowledged insecure tenure rights as a driver of deforestation, and 31 identified the clarification of tenure as key to the implementation of REDD+ (RRI, 2012a).

However, the steady increase in commodity prices resulting from demand through the 2000s for biofuels, food, and other raw materials found and produced in forest areas also drove the expansion of industrial concessions. In 2012, the International Land Coalition (ILC) identified up to 203 Mha in land acquisitions approved or under negotiation between 2000 and 2010, with the rate of acquisition increasing dramatically between 2005 and 2009 (Anseeuw *et al.*, 2012, 4). Between October 1, 2008 and August 31, 2009 alone, Deininger *et al.* (2011) identified 71 Mha allocated globally for large-scale land acquisitions. While these estimates are contested,[14] they provide an indication of the dimension of large-scale land acquisitions over this time period. Furthermore, given that the large-scale transfer and acquisition of lands by companies is not a recent feature of the tenure landscape, these time-bound estimates do not fully capture the global footprint of concession agreements.

One aspect of these investments is quite apparent – low- and middle-income forest countries are a major destination (RRI, 2013a, 2013b, Forthcoming) for investments in oil palm (ABN, 2007; Casson, 2003; Schoneveld *et al.*, 2010; Venter *et al.*, 2009), logging – in spite of moratoria (Greenpeace, 2012; Global Witness, 2013; Forest Trends and Rights and Resources Initiative, 2013), and oil and gas exploration (Flórez, 2013; Espinoza and Feather, 2011), amongst other uses. Investors include major multinational companies, capitalized in the hundreds of billions, with supply chains stretching across multiple countries, regions and continents. More numerous though, are the small- and medium-sized firms, whose supply chains and revenue streams are typically geographically limited to a particular country or region.

Governments are attempting to capitalize on the growing demand for natural resources by ceding the right to develop domestic natural resources to third parties in exchange for a stream of payments or other benefits. They have seen these concessions as a means to decrease dependence on aid, generate formal employment, and increase national incomes. However, in the rush to open their countries for investments, governments have repeatedly sidelined communities in negotiations – both those who have legal rights to own and control their lands, and those who do not.

The impacts of contestation

The body of literature on the impacts of large-scale land acquisitions that neglect local rights is extensive – and growing. These impacts range from the forcible eviction and destruction of homes (ABN, 2007; Schoneveld *et al.*, 2010), to the destruction of sacred sites (Afrol News, 2011; FOEI, 2013); the appropriation of lands and destruction of resources that are essential for local livelihoods (FOEI, 2013; Overbeeke, 2010), to social and political conflict (Monachon and Gonda, 2011; Zander and Dürr, 2011).

The mechanisms through which individuals and communities resist these encroachments are also widely documented. Common approaches include protests, complaints to local officials, legal challenges, allying with NGOs to launch awareness campaigns, and/or the use of sabotage or violence – all of which create delays and increase operational costs for the concessionaries. Communities and their civil society advocates are also beginning to use dispute-resolution mechanisms in product-certification bodies like the Roundtable for Sustainable Palm Oil (RSPO) and FSC to challenge large multinational corporations, often in countries where the court systems lack the capacity to effectively hear a case or can be corrupted by more powerful interests. And while the outcomes of these challenges are not legally binding, they can inflict substantial reputational costs on the corporation, as well as compel companies to reexamine their business practices, or lose valuable certifications. One issue which is beginning to be explored in the literature is the costs for investors who fail to conduct proper due diligence on the tenure implications of their operations. These are costs to which the private sector is beginning to awaken.

A 2012 study found that companies that neglect local rights – even those that are not necessarily recognized in law or formally delimited – can incur substantial costs, through operational delays, legal costs, and counterparty and reputational risks (Munden Project, 2012). With sufficient mobilization, local opposition can threaten the viability of an investment; affect the credit-rating of the company; and even cause the financial collapse of the project or the withdrawal of support by the government or by primary investors. In the example of Stora Enso, one of the world's largest pulp and paper companies, the firm's land acquisition practices in southern China resulted in local protests and violent conflict (Ping and Nielsen, 2010). Conflict between the company and local farmers was driven by the company's failure to perform adequate due diligence for their acquisitions. The company relied heavily on local government and middlemen in the transactions, rather than directly negotiating with communities and fully ensuring the legality of their transactions. The 2010 study demonstrated that local landholders were neither properly consulted nor properly compensated, and in some cases were forcibly coerced into transferring their land. Following the backlash from these incidents, which resulted in major operational delays, increased costs, and reputational damage, Stora Enso implemented a moratorium on leasing any further collectively owned lands, and have been reviewing their acquisition policies and working with civil society organizations in an attempt to rectify the tenure issues surrounding their investments and avoid future conflicts, though it appears that the company still has some steps to take to ensure that all their lands have been legally acquired (Ping and Xiaobei, 2014).

In the oil palm sector, which is highly active in the world's forest areas, Indigenous Peoples and local communities have increasingly begun to use fora such as the RSPO to confront companies about their land-rights-based abuses. A preliminary review of the cases submitted to the Dispute Resolution Facility of the RSPO, revealed that of the 40 cases received against palm oil growers and processors since the mechanism was put into place in 2009, 23 were explicitly based on community grievances, including violations of or disputes over land rights, intimidation of community members during consultation, failures to fully implement FPIC, contamination of community lands or water as a result of the processes used on the plantations, and the use of violence against protesters (RSPO, 2014). The experience of the Malaysian firm Sime Darby, the world's largest producer of certified palm oil, is a useful example of the dispute resolution function of industry roundtables. In 2009, the company signed a 63-year concession with the Government of Liberia for 220,000 hectares of land on which to develop oil palm and rubber. However, in 2011, villagers complained that in initial operations, Sime Darby had forced them off their customarily held lands, cleared forests, and destroyed wetlands that were essential to their local livelihoods (RRI, 2012b). With the assistance of civil society organizations, communities formally raised the issue with the RSPO, which resulted in suspension of company operations pending an agreement and bilateral discussions between the company and community members. The delays resulting from the dispute have hampered the company's efforts to scale up their production in the country.

The mining sector, a significant driver of deforestation, is also vulnerable to the costs of disruption resulting from tenure risk (Sosa, 2011). According to the UN Special Representative for Business and Human Rights, operational disruptions resulting from community protests can cost global mining operations between US$20 million and US$30 million a week (Business Ethics, 2011).

The physical extent of 'tenure risk' for investors who neglect the importance of recognizing local communities' claims to the resources is not to be underestimated. One study that assessed the overlap between existing industrial concessions and customarily claimed lands gathered geo-spatial data on lands claimed, controlled, or owned by Indigenous Peoples and local communities and forest, mineral, and agricultural concessions in 12 countries in South America, sub-Saharan Africa, and Southeast Asia (Munden Project, 2013). Of the 153.5 Mha of concessions examined, at least 31% overlapped with community-held lands in some way. Given the difficulties in accessing reliable data (on both concessions and community-claimed land), this figure very clearly underestimates the true extent of the overlap.

Another study by First Peoples Worldwide examined the activities of the energy and mining companies listed in the Russell 1,000 Index and their impacts on Indigenous Peoples' forest and other lands (First Peoples Worldwide, 2013). It found that over 30% of current oil and gas production is currently sourced on or near Indigenous Peoples' lands, which also account for nearly 50% of known oil and gas reserves. The study found that over 40% of current mineral production is sourced on or near indigenous lands, as will be nearly 80% of known future projects. Furthermore, the study found that 92% of the sites included in the study posed a medium to high risk to shareholders.

As the First Peoples Worldwide study demonstrates, the absence of clarity about forest tenure rights is not a problem that exclusively affects low- and middle-income countries. Many of the mining sites identified in the study as most vulnerable to contested tenure were in Canada. This finding is corroborated in a study by the Fraser Institute (2013) which revealed that disputed land claims with aboriginal communities and individuals had become a major factor deterring mining investment in British Columbia, Canada. Contested tenure is also an issue in the forest sector in Canada. Moreover, in December 2013, the FSC suspended some certifications from Canada's largest forestry company, Resolute Forest Products, following a complaint by the Grand Council of the Crees that the company had not followed FPIC during their activities (Forest Stewardship Council, 2013).

These findings make it increasingly apparent that government agencies and investors in land and forest resources will need to evolve beyond traditional models of extraction and exclusion, by developing more sophisticated understandings of land and natural resource tenure, and develop mechanisms to equitably engage with the Indigenous Peoples and local communities living there in ways that respect their rights. The failure to develop these mechanisms and tools can only lead to the continued destruction of the world's remaining forest resources and the devastation of the lives of those whose livelihoods and identities depend on them.

The future of forest reform?

The substantial amount of land in forest areas owned and controlled by Indigenous Peoples and local communities represents significant social and political progress. It increases the chances for cultural survival and locally determined development; justly positions forest communities as key actors in local, national, and global forest management; and further highlights the need for their participation in the ongoing global discourse on conservation and climate mitigation efforts. The increasing power of indigenous and community organizations, through the global networking of their efforts and through the development of new tools to assert and defend their rights, is good news for forests.

At the same time, demands on forests and the resources they contain continue to rise, and Indigenous Peoples' and local communities' tenure rights remain only partially realized. The limited protection for existing rights, and remaining vast extent of state claims to forests, and the documented slowdown in the recognition of rights are significant barriers that continue to hamper the development of more inclusive and equitable mechanisms of engagement between communities and those who wish to invest in forest lands or to ensure their conservation. How is the world to reduce pressure on forest areas when some of the key drivers of deforestation – the lack of clear tenure rights, and tenure conflict – have not yet been addressed?

Many cost-effective methods exist for better securing local tenure rights that combine formal survey, titling and registration activities, adjudication, strengthening of customary resource governance, and recognition of collective boundaries. However, in many countries now planning or engaged in land reforms, these methods and best practices are often not known or put into practice.

The world cannot afford to continue to ignore the essential question of who owns and controls forest lands and resources. It is clear that insecure tenure and tenure conflict are drivers of deforestation and poverty, and inhibit sustainable and equitable investment. Indigenous and community tenure rights need to be at the center of local, national, and international policy debates and decisions about forest land and resources. This will require both increased support for the self-determined initiatives of Indigenous Peoples and local communities, and consistent respect and protection of rights by governments, business, investors, and other actors concerned with the future of forests.

Notes

1 Each of these rights – management, exclusion, unlimited duration, and due process and compensation are fully defined in RRI (2014).
2 Countries in order of total forest area: Russian Federation, Brazil, Canada, United States, China, Democratic Republic of Congo, Australia, Indonesia, India, Peru, Mexico, Colombia, Angola, Bolivia, Zambia, Tanzania, Myanmar, Papua New Guinea, Japan, Central African Republic, Republic of the Congo, Finland, Gabon, Cameroon, Thailand, Lao People's Democratic Republic (PDR), Guyana, Philippines, Suriname, Vietnam, Ethiopia, Cambodia, Honduras, Republic of Korea, Nepal, Kenya, Bhutan, Costa Rica, Gambia, and Togo. An additional 12 countries were listed in the report, but did not have complete data.
3 This study identifies low- and middle-income countries as those having a gross national income (GNI) per capita lower than US$12,616, as ranked by the World Bank. http://data.worldbank.org/about/country-classifications (accessed 12 Dec. 2013).
4 In descending order by forest area: Russian Federation, Brazil, Canada, United States, China, Democratic Republic of the Congo, Australia, Indonesia.
5 These nine countries are Angola, Gabon, Cameroon, Thailand, Lao PDR, Guyana, Vietnam, Cambodia, and Honduras.
6 Most of these countries have national-level legislation to recognize community rights, but have either failed to implement the tenure regimes, or the rights recognized within those regimes are insufficient to exert any meaningfully legal control over their resources.
7 In Mozambican law, all lands belong to the state. However, the communities have sufficient legal rights to constitute 'ownership' within the parameters of this study.
8 The Mozambican government reports the total demarcated and delimited community lands; however, under law, communities can enjoy ownership rights to their lands without demarcation – therefore, using official figures would greatly under-represent the total area legally under community ownership.
9 The Congo Basin countries represented within this study include: Angola, Cameroon, the Central African Republic, Democratic Republic of Congo, Gabon, and the Republic of the Congo.
10 The only country in sub-Saharan Africa to have ratified this is the Central African Republic.
11 Papua New Guinea accounts for much of the balance of forest lands owned by Indigenous Peoples and other communities in Asia, with nearly 18% of the total regional share in 2013. The Philippines and India are the only other two countries identified in the region with implemented tenure regimes that recognize community ownership of lands.

12 Mina Setra. The Indigenous Peoples' Alliance of the Archipelago (2013). Personal communication.
13 Some of this area may be non-forest area.
14 These estimates have been criticized due to the methodology used in compiling these data. The ILC has recently revised the total confirmed cases to cover over 35.6 Mha, with an additional 14.1 Mha as "intended" investments (Land Matrix data as accessed 10 February, 2014). However, a series of independent studies by RRI in Liberia, Cameroon, Lao PDR, Peru, and Myanmar used a methodology primarily based on assessing company financial reports and using government concessions data sets, and have been able to confirm a much higher level of investments than are currently reported for these countries by the ILC's Land Matrix. Furthermore, these agreements are often negotiated in contexts of limited transparency, and several companies which are not publically traded do not publish their financial statements, making it all the more difficult to assess the true extent of investments. These early estimates should therefore be viewed as effective illustrations of the scale of the issue over the past decade alone, but not necessarily as precise figures.

References

ABN. 2007. *Agrofuels in Africa: The impacts on land, food and forest.* African Biodiversity Network.

Afrol News. 2011. Ethiopian "sacred forests" sold to Indian tea producer. 18 February, 2011. http://afrol.com/articles/37365.

Alden Wily, L. 2012. *Customary Land Tenure in the Modern World, Rights to Resources in Crisis: Reviewing the fate of customary tenure in Africa.* Brief 1 of 5. Washington, DC: Rights and Resources Initiative.

Anseeuw, W.*et al.* 2012. *Land Rights and the Rush for Land: Findings of the global commercial pressures on land research project.* Rome: International Land Coalition.

Buhmann, K and I. Nathan. 2012. Plentiful forests, happy people? The EU's FLEGT approach and its impact on human rights and private forestry sustainability schemes. *Nordic Environmental Law Journal*, 4, 53–82.

Business Ethics. 2011. Business and Human Rights: Interview with John Ruggie. 30 October, 2011. http://business-ethics.com/2011/10/30/8127-un-principles-on-business-and-human-rights-interview-with-john-ruggie.

Carniero, J. 2014. Brazilian former slave community fights for land. BBC Brasil. 7 January, 2014. www.bbc.com/news/world-latin-america-25622027.

Casson, A. 2003. *Oil Palm, Soybeans and Critical Habitat Loss.* Hohlstrasse, Switzerland: World Wildlife Fund Forest Conversion Initiative.

Chhatre, A. and A. Agrawal. 2009. Trade-offs and synergies between carbon storage and livelihood benefits from forest commons. *PNAS*, 106(42): 17667–17670.

Deininger, K.*et al.* 2011. *Rising Global Interest in Farmland: Can it yield sustainable and equitable benefits?* Washington, DC: The World Bank.

Espinoza, L.R. and C. Feather. 2011. *The Reality of REDD+ in Peru: Between theory and practice.* Lima, Peru: Forest Peoples Programme (FPP), Central Ashaninka del Río Ene (CARE), Federación Nacional Nativa del Río Madre de Dios y sus Afluentes (FENAMAD) and Asociación Interétnica de Desarrollo de la Selva Peruana (AIDESEP).

FERN. 2013. *Improving Forest Governance: A comparison of FLEGT-VPAs and their impact.* Brussels: FERN.

First Peoples Worldwide. 2013. *Indigenous Rights Risk Report for the Extractive Industry (U.S.): Preliminary findings*. Fredericksburg, USA: First Peoples Worldwide.

Flórez, M. 2013. *Impacto de las Industrias Extractivas en los Derechos Colectivos sobre Territorios y Bosques de los Pueblos y las Comunidades*. Rights and Resources Initiative, Asociación Ambiente y Sociedad.

FOEI. 2013. *Sime Darby and Land Grabs in Liberia*. Amsterdam: Friends of the Earth International.

Forest Stewardship Council. 2013. *Suspension of Resolute FSC Certificates*. December 13, 2013. Accessed March 2014. https://ic.fsc.org/newsroom.9.605.htm.

Forest Trends and Rights and Resources Initiative. 2013 (Unpublished). *Conversion Timber in FLEGT countries*.

Fraser Institute. 2013. *British Columbia's Mining Performance: Improving BC's attractiveness to mining investment*. Vancouver, Canada: Fraser Institute.

Global Witness. 2013. *Avoiding the Riptide: Liberia must enforce its forest laws to prevent a new wave of illegal and destructive logging contracts*. London, UK: Global Witness.

Greenpeace. 2012. *Up for Grabs: Millions of hectares of customary land in PNG stolen for logging*. Ultimo, Australia: Greenpeace.

Larson, A.M. *et al*. 2010. Rights to forests and carbon under REDD+ initiatives in Latin America. *CIFOR Infobrief*, 33: 1–8. www.cifor.org/library/3277/rights-to-forests-and-carbon-under-redd-initiatives-in-latin-america.

Monachon, D. and N. Gonda. 2011. *Liberalization of Ownership versus Indigenous Territories in the North of Nicaragua: The case of the Chorotegas*. Washington, DC: International Land Coalition.

Munden Project. 2012. *The Financial Risks of Insecure Tenure: An investment view*. Washington, DC: The Rights and Resources Initiative. www.rightsandresources.org/documents/files/doc_5715.pdf.

Munden Project. 2013. *Global Capital, Local Concessions: A data driven examination of land tenure risk and industrial concessions in emerging market economies*. Washington, DC: The Rights and Resources Initiative. www.rightsandresources.org/documents/files/doc_6301.pdf.

Nolte, C. *et al*. 2013. Governance regime and location influence avoided deforestation success of protected areas in the Brazilian Amazon . *PNAS*, 110(13): 4956–4961.

Oakland Institute. 2013. *On Our Land: Modern land grabs reversing independence in Papua New Guinea*. Oakland, USA: Oakland Institute.

Okereke, C. and K. Dooley. 2010. Principles of justice in proposals and policy approaches to avoided deforestation: towards a post-Kyoto climate agreement. *Global Environmental Change*, 20(1), 82–95.

Overbeeke, W. 2010. *The Expansion of Tree Monocultures in Mozambique. Impacts on Local Peasant Communities in the Province of Niassa*. Montevideo, Uruguay: World Rainforest Movement. wrm.org.uy/books-and-briefings/the-expansion-of-tree-monocultures-in-mozambique-impacts-on-local-peasants-communities-in-the-province-of-niassa.

Ping, L. and R. Nielsen. 2010. *A Case Study on Large-Scale Forestland Acquisition in China: The Stora Enso Plantation Project in Hepu County, Guangxi, Province*. Washington, DC: Rights and Resources Initiative, and the Rural Development Institute.

Ping, L. and W. Xiaobei. 2014. *Forestland Acquisition by Stora Enso in Southern China: Status, issues and recommendations*. Washington, DC: Rights and Resources Initiative.

Porter-Bolland, L.*et al*. 2012. Community-managed protected areas: An assessment of their conservation effectiveness across the tropics. *For. Ecol. Manag.*, 268, 6–17.

REDD-Monitor. 2014. *Peru approves the expansion of the Camisea gas project into indigenous peoples' reserve.* www.redd-monitor.org/2014/01/30/peru-approves-the-expansion-of-the-camisea-gas-project-into-indigenous-peoples-reserve.

RRI. 2012a (unpublished). *Internal assessment of national R-PP and R-PIN documents.* Washington, DC: Rights and Resources Initiative.

RRI. 2012b. *Turning Point: What future for forest peoples and resources in the emerging world order?* Washington, DC: Rights and Resources Initiative.

RRI. 2013a. *Investments into the Agribusiness, Extractive, and Infrastructure Sectors of Liberia, An overview.* Washington, DC: Rights and Resources Initiative.

RRI. 2013b. *Investments into the Agribusiness, Extractive, and Infrastructure Sectors of Cameroon, An Overview.* Washington, DC: Rights and Resources Initiative.

RRI. 2014. *What Future for Reform? Progress and slowdown in forest tenure reform since 2002.* Washington, DC: Rights and Resources Initiative.

RRI. Forthcoming. *Investments into the Agribusiness, Extractive, and Infrastructure Sectors of Peru, Lao PDR, and Myanmar.* Washington, DC: Rights and Resources Initiative.

RSPO. 2014. Status of Complaint. Accessed 26 Feb. 2014. www.rspo.org/en/status_of_complaint.

Schoneveld, G.C.*et al*. 2010. *The Role of National Governance Systems in Biofuel Development: A comparative analysis of lessons learned.* Bogor, Indonesia: Center for International Forestry Research.

Sosa, I. 2011. *License to Operate: Indigenous relations and free, prior and informed consent in the mining industry.* Amsterdam: Sustainalytics.

Sunderlin, W.*et al*. 2008. *From Exclusion to Ownership? Challenges and opportunities in advancing forest tenure reform.* Washington, DC: Rights and Resources Initiative.

Venter, O.*et al*. 2009. Carbon payments as a safeguard for threatened tropical mammals. *Conservation Letters*, 2, 123–129.

White, A. and A. Martin 2002. *Who Owns the World's Forests? Forest tenure and public forests in transition.* Washington, DC: Forest Trends and Center for International Environmental Law.

Zander, M. and J. Dürr. 2011. *Dynamics in land tenure, local power and the peasant economy: The case of Petén, Guatemala.* Paper presented at the International Conference on Global Land Grabbing, 6–8 April. Organized by the Land Deals Politics Initiative.

7 Enabling investment for locally controlled forestry

Duncan Macqueen and Peter deMarsh

Why locally controlled forestry matters

Aligning business (and development more broadly) with sustainable management of the world's forests is a pressing challenge. Forests cover almost one third of the world's land area and are critical for mitigating global climate change, producing food, fuel, fibre and a host of other products, and maintaining the integrity of bio-cultural ecosystems. Nearly all forests are inhabited – by approximately 1.3 billion families, communities and indigenous peoples – most of whom are forest and farm producers,[1] attuned to their forest and farm landscape (Chao, 2012). But as we note below the degree of locally controlled forestry varies (see definition in the second section of this chapter). Locally controlled forestry also encompasses different ownership and management within family smallholdings, community forestry and indigenous people's territories, and different scales of operation. Because of this variety, care is needed in attributing specific traits to locally controlled forestry – with different authors making assessments of different subcomponents of that broader whole (e.g. looking only at community forestry management, or only at small and medium forest enterprises). We try to make explicit to what different authors are referring in the referencing within this chapter in order to avoid any charge of revisionism.

Globally, the forest area under legal community ownership or control – including both indigenous and non-indigenous community forestry – has risen from approximately 11% in 2002 to 15.5% in 2013. In low- and middle-income countries, the forest area under legal community ownership or control has risen from over 21% in 2002 to over 30% in 2013 (RRI, 2014). There are also many more forest and farm producers living in or adjacent to forests without formal control over forest resources. The good economic, social and environmental reasons for increasing the degree and formality of local control over forests are described in the next chapter. Yet, the sheer scale of forest and farm producers is at once an opportunity and a challenge.

As an *opportunity*, myriad locally controlled forest enterprises constitute a vast forest-related private sector in which benefits to livelihoods and forest condition go hand in hand. Landscape-scale improvements in forest condition have emerged in countries where the potential of the community subcomponent of locally controlled forestry has been unleashed, such as Mexico, Nepal and Tanzania (Seymour *et al.*, 2014). This community-managed forestry subcomponent of locally controlled forestry has been shown to be at least as effective as state-enforced protected areas as a means of stemming forest loss (Porter-Bolland *et al.*, 2012), and that it has generally positive impacts on forest condition (Bowler *et al.*, 2010). Contrary to common perception therefore, forest protection and economic development can be not just compatible, but complementary, without difficult trade-offs, at least within the subset of forestry that is locally controlled. Improved condition in locally controlled forests has been accompanied by substantial livelihood benefits across a wide range of contexts – including family small-holdings (Ackzell, 2009), community forests (Bray *et al.*, 2003; Molnar *et al.*, 2007; Charnley and Poe, 2007; Ojha *et al.*, 2009) and Indigenous Peoples' territories (Nepstad *et al.*, 2006; CEESP, 2008). This is in fairly sharp contrast with the dominant subset of forestry controlled by large-scale industries, in which little alignment was found between commercial activity and either forest protection or poverty reduction (Mayers, 2006a).

Given the strong links between locally controlled forestry and both forest protection and poverty reduction it is worth interrogating further how these links are brought about. Income from locally controlled forestry accrues locally and can be reinvested locally, thereby providing an incentive to sustain forests for the multiple benefits that are best appreciated locally. This strengthens local capacity for sustainable business and helps build business organisations that champion the multiple benefits of forests in ways that respect cultural norms (Macqueen, 2007). There is a strong case for locally controlled forestry being the best bet for aligning development and business with sustainable management of the world's forests.

Forest and farm producers are embedded within, and have an integrated outlook regarding, forest and farm landscapes. They are well aware of the need to maintain their own livelihoods through community cohesion, income generation, food security, energy provision, and so on – referred to as social foundations in the broader analysis of Raworth (2012). They are well aware of the local need to maintain the integrity of ecosystems upon which their survival depends, locally adapting to and mitigating climate change, conserving necessary biodiversity, and as farmers, maintaining long-term soil fertility without costly and potentially resource-degrading chemical inputs, and especially, maintaining water tables and flows of water from the forests both for drinking water and farming requirements. Their daily experience provides an ongoing demonstration of how the environmental, social and economic dimensions of their lives are a single, integrated reality.

This local integration and integrity matters because without it it is difficult to see how production at a global level could avoid breaching planetary boundaries (see Rockström *et al.*, 2009). Attempts to secure both local integrity and planetary level integrity – productive use that does not breach planetary boundaries – are much more likely to succeed through the multi-functional mosaics of locally controlled forestry, than through the monotypic expanses of large-scale industrial forestry (Macqueen, 2013a).

As a *challenge*, providing the necessary support to unleash the potential of locally controlled forestry over extensive areas, with variable human capacity and infrastructure, is formidable. There is relative neglect of forest and farm producers in government and development agency policies and programmes. Perceptions prevail of the impossible complexity of coordinating the management of the vast number of these forest parcels in a productive way, and as a corollary, the unlikelihood that these families and communities have the capacity to develop effective organisations to provide necessary support for proper management (deMarsh *et al.*, 2014). Large-scale, industrial forestry is treated as the norm. It is preferred because of its capacity to focus narrowly on economic returns from one or a small number of products. It is preferred because the large corporations who carry it out have access to significant financing, and are able to make large investments that entail large-scale and high-profile employment creation, using the latest technology. The low transaction costs in its administration and revenue collection are also considered an advantage. Yet in reality the bargaining power of large-scale, industrial forestry often makes these revenue advantages illusory (see Bulkan, 2014).

The contrast with locally controlled forestry is dramatic: multiple products many of which never enter conventional markets, scattered part-time employment, extremely limited access to capital, rudimentary technology and high transaction costs for any formal activities that are engaged in. Each unit is so small, that data collection on locally controlled forestry is neglected. Formal activities become almost invisible from the perspective of this dominant paradigm, despite their huge aggregate significance. Estimates of the large aggregate significance of small and medium forest enterprises, for example, suggest that they make up 80–90% of enterprise numbers and more than 50% of employment in many countries (Mayers, 2006b). In addition, the subsistence and informal economic components of locally controlled forestry are not recognised by the formal economy, leading to further under-representation in official statistics.

How investing in locally controlled forestry can be brought about

In recognition both of the opportunity and the challenge of locally controlled forestry that had been highlighted in previous years (see Mayers, 2006b) representatives of family, community and indigenous forestry began a discussion process in 2009 to work out a common agenda (via leaders of the International Family Forestry Alliance, the Global Alliance for Community Forestry and the International Alliance of Indigenous and Tribal Peoples of the Tropical Forests – referred to as 'The Three Rights-holders Groups', abbreviated to 'The G3'). While representation by such international alliances is inevitably partial, the process of meeting did for the first time develop a credible, combined position, agreed by all parties as to the terminology and agenda that they wished jointly to pursue (G3, 2011). They agreed to work around possible areas of argument or fracture lines to do with: indigenous rights versus non-indigenous interests; collective versus individual rights; land use conflicts; and differences in methods of economic, financial and forest land management.

The G3 adopted the mutually acceptable terminology 'locally controlled forestry' as a shorthand for their joint agenda and defined it as:

> The local right for forest owner families and communities to make decisions on commercial forest management and land use, with secure tenure rights, freedom of association and access to markets and technology.

Between 2009 and 2012, a series of 11 international dialogues took place between representatives of family, community and indigenous forestry, investors and forest experts under the title 'Investing in Locally Controlled Forestry (ILCF). Pre-conditions for successful locally controlled forestry, and practical strategies for mobilising greater investment in support of this agenda were examined (Macqueen *et al.*, 2012a). It was recognised that achieving compatibility between rights-holders, investors and governments requires an adjustment of the conventional investment approach, characterised as 'capital seeks forest resources and needs labour'. A new paradigm was advanced, which can be stated as 'rights-holders manage forest resources and seek capital' (Elson, 2013). It was also recognised that two types of investment are essential for successful locally controlled forestry: *enabling investment* in rights, organisation and technical and management capacity, from which a tangible financial return is not expected – but which creates the conditions for: *asset investment* in improved forest management, processing facilities and marketing capacity from which a tangible financial return is expected. The conditions established by the former serve to attract the latter – but on a fairer and more sustainable footing than has usually been the case.

Forest and farm producers from varied contexts undertake a range of subsistence, informal and formal business activities. Recent evidence suggests that no single activity can provide the full range of local and global public goods required by humanity from forest and farm landscapes (Macqueen *et al.*, 2014a). Approaches to enabling locally controlled forestry must therefore be found that are relevant to a wide variety of forest businesses – including both products and service subsectors for which an indicative typology is given in Table 7.1. The value associated with the first three categories of products in this typology makes them particularly important to consider in what follows (see for example: Nhancale *et al.*, 2009; Gebremariam *et al.*, 2009; Osei-Tutu *et al.*, 2010).

Through the discussion processes and dialogues described above, it has become clear that notwithstanding differences in context and product or

Table 7.1 Typology of product and service subsectors that can form part of the portfolio of locally controlled forest enterprises

Subsectors	*Secondary divisions*	*Examples*
1. Biomass energy	Fuel wood	Firewood branches and chopped logs
	Charcoal	Rough charcoal or compacted charcoal briquettes
	Wood pellets	Chipped wood that may be dried to differing degrees
2. Industrial round wood	Logs	Sawn logs that may or may not be debarked
	Pulpwood	Sawn logs (including smaller dimension stems and branches)
3. Primary processed products	Sawn wood	Planks and posts
	Pulp for paper	Pulp feedstock
	Paper products	Paper and paper board
4. Secondary processed products	Furniture and parts	Wooden chairs, office, kitchen or bedroom items
	Builder's joinery or carpentry	Wood panels, parquet panels, shingles and shakes
	Shaped wood	Unassembled parquet, strips, friezes, tongued, grooved, beaded, moulded and rounded wood
5. Non-timber forest products	Food products	Fruits, nuts, seeds, including coffee and honey
	Oils and resins	Woodworking oils, cosmetic and medicinal oils, resins and gums
	Fibre products	Thatch, wickerwork furniture, craft use
	Ornamental plants	Flowers, houseplants, garden and urban amenity planting
	Medicinal plants	Various remedies for internal and external application
6. Services	Tourism	Parks, recreational sites (hiking, biking, canopy adventures, etc.)
	Biodiversity conservation	Forest protection and management
	Watershed protection	Riparian strips, cover and steep slopes, etc.
	Carbon sequestration	Forest management and restoration
	Hunting and gathering	Licensed harvesting

service, the enabling conditions for sustainable business are very much the same, in all parts of the world (see for example Macqueen, 2013b). These enabling conditions can be expressed through simple questions that any forest and farm producer might ask before embarking on investment in a formal (or indeed, informal) business (see Table 7.2).

Each of these questions reflects a particular type of risk with which producers have experience. Forest and farm producers are no different from any other type of investor: before undertaking an investment in a new practice or product, they will assess the risks of which they are aware. The fact that they survive often incredibly adverse circumstances suggests they do it quite well. They differ from more wealthy investors in the extent to which threats to basic subsistence needs override apparent opportunities to increase incomes. A combined package of enabling investments that helps forest and farm producers to answer 'yes' to each and every question in the list above is likely to result in successful locally controlled forestry that aligns business with protection of the world's forest. It should be noted that all four conditions need to be satisfied before a forest and farm producer would embark on a particular business venture – they must be delivered as a package. None are optional extras. We discuss each in turn below.

Enabling condition 1 – Secure tenure

Forests are often a valuable and therefore contested resource. It is quite normal for a range of different potential claimants to put forward arguments in favour of their rights. Decisions about forest and farm tenure – who can use what resources, over how long and under what conditions – are critical

Table 7.2 Enabling conditions and descriptive questions for locally controlled forestry

Enabling condition	*Descriptive question to which a forest farm producer must be able to answer yes if they are to plant or manage trees*
Secure commercial tenure	Question 1. If I plant tree *x* (for food, fuel, fibre, conservation, etc.) will I have the right to sell it?
Fair market access and business support	Question 2. If I plant tree *x* (for food, fuel, fibre, conservation, etc.) will I be able to sell it at a fair price?
Appropriate technical extension support	Question 3. If I plant tree *x* (for food, fuel, fibre, conservation, etc.) will I be able to get the management and technical support to manage it sustainably, protect it from pests and diseases and package it for the market?
Freedom of association/ strength of organisation	Question 4. If I plant tree *x* (for food, fuel, fibre, conservation, etc.) will I be able to associate with others to make sure circumstances don't change while the trees are growing and I can carry on answering yes to the previous three questions?

(Mayers *et al.*, 2013). They determine what forest business options exist – and as noted above, are often rigged in favour of large-scale industrial forestry. Where forest and farm producers have lived in or by forests for generations – often long before the delineation of national boundaries – there is a moral imperative to weight tenure in their favour, and indeed, to build on their customary management arrangements (Gilmour and Fisher, 2011). But it also makes economic sense. Forestry is a long-term business. Planting trees is hard work and costs money. Nobody does it (particularly not poor risk-averse forest and farm producers) unless they can be confident that selling the trees will benefit themselves, their children or their grandchildren.

Ensuring secure forest tenure is just as important to locally controlled forestry as it is for any large-scale industrial forestry. A number of global and cross-regional studies affirm the importance of tenure – in a range of family, community and indigenous contexts – to deliver both good forest condition and livelihood benefits (Pagdee *et al.*, 2006; Lawry *et al.*, 2011; Persha *et al.*, 2011; Robinson *et al.*, 2011; Macqueen, 2013b).

Locally controlled forestry requires confidence in a number of areas before embarking on a business (see the particular contextual treatment of RRI and ITTO, 2009):

- *Duration* – Rights must last long enough to make investing in sustainable forest business worthwhile
- *Assurance* – Rights must guarantee that forest and farm producers benefit from the returns or their investment free from interference
- *Robustness* – Rights must be enforced and easy to defend in a court of law
- *Exclusivity* – Rights must in no way overlap with the rights of external investors or government agencies
- *Simplicity* – Rights must be simple to acquire and free of excessive bureaucracy or costly registration

It is therefore not just the allocation of rights that is important but the way in which those rights (and associated obligations) are prescribed, administered, supported and enforced – good forest governance – that makes locally controlled forestry flourish. In many cases where local forest and farm producers are given formal forest management rights their utility is often limited by associated obligations or by other state rules and regulations (Larson and Dahal, 2013).

Good governance of forest tenure, and indeed also tenure beyond forests, involves a number of key ingredients (FAO, 2012): respecting legitimate tenure rights and the people who hold them, safeguarding those rights against threats, promoting and facilitating the enjoyment of those rights, providing access to justice to deal with infringements, and preventing conflicts and opportunities for corruption.

In the light of the global forest situation, there are good reasons, and in many cases also good prospects for pursuing tenure reform. How to do so in practice is the subject of a practical guide to improving the governance of forest tenure which usefully draws together, and helps illustrate how to use numerous tools and tactics (Mayers *et al.*, 2013). The guide categorises tools around four stages of possible engagement:

- *Tools for understanding* – how to assess the historical and current resource dynamics, tenure and governance contexts
- *Tools for organising* – how to develop the legitimacy, accountability and effectiveness of groups and institutions so as to have policy influence
- *Tools for engaging* – how to engage with the processes of decision-making that effect tenure (on the spectrum from cooperative dialogue to resistance)
- *Tools for ensuring* – how to develop mechanisms for accountability to make sure that dialogue and promises translate into action

Locating an appropriate starting point by assessing the current situation in-country is the preliminary step in making use of such tools.

Enabling condition 2 – Fair market access and business support

Fair market access and business support is also fundamental to successful locally controlled forestry (deMarsh *et al.*, 2014). Investing in improved forest production only makes sense if there is a reasonable prospect of selling the eventual production on acceptable terms. And turning those prospects into reality often requires business capacity development. It may also require scaling-up (in order to gain cost efficiencies on inputs and better sales margins through more powerful negotiations with buyers) – an issue which is dealt with under Enabling condition 4 below.

Subsistence use rights (to biomass energy, timber and non-timber forest products – NTFPs), or market access for low-value NTFPs, or allocations of forest land that are heavily degraded in terms of timber are often readily granted by governments to local groups or individuals. Yet, because of their commercial value, governments are often reluctant to grant local forest and farm producers commercial sales rights for timber and biomass energy (Scherr *et al.*, 2003; RRI, 2012; Larson and Dahal, 2013).

In terms of timber, in their global review, RRI (2012) found that some 17% of countries surveyed (nine out of a total of 52) explicitly prohibit communities from commercially selling timber and the remainder (43 out of a total of 52) must comply with management plans and/or licenses, many of which pose insuperable challenges for local forest and farm producers. This contrasts with NTFPs where 90% of all surveyed regimes (53 out of a total of 59) recognise community rights to harvest NTFPs, but, of those, 11% (six out of a total of 53) restrict use to subsistence.

A few examples are illustrative of the problems encountered. In India in some states, communities under joint forest management arrangements not only have to apply for up to 10 different permits to sell timber, but they must also share commercial benefits with Government (Long, 2010). In Mozambique communities with strong rights under the 1997 Land Law are constrained in making commercial sales of timber, that are theoretically available under simple license and concession regimes, by the onerous administrative, management planning, and in the case of concessions, value added processing investment required under the 1999 Forest Law (Nhancale *et al.*, 2009). In Myanmar, where the 1995 Community Forestry Instructions laid out the possibility of community forest user groups delimiting forest land and selling surplus product, the sale of teak (the commercially most valuable species) was explicitly prohibited (Tint *et al.*, 2014). In Nepal, although the Forest Act and Rules grant communities rights to sell timber, in practice the complex formalities and procedures of management plan development, permits to transport and trade timber from private forests, and the requirement to meet district level demand before trading beyond it, impede access to timber markets (ANSAB, 2009).

In terms of market value, biomass energy may even exceed that of timber for many developing countries, especially in Africa and South or South East Asia where it makes up more than 60% of household energy consumption. The sector is plagued by high levels of informality – fuelled by often unsubstantiated assumptions that biomass use inevitably leads to unsustainable forest management or increases household respiratory diseases, and equally importantly, by the knowledge that huge rent-seeking opportunities exist from its criminalisation. Such is the negative perception of biomass energy that many government agencies do not even collect statistics on it (Openshaw, 2010). For the reasons outlined above, governments tend to place stringent restrictions on market access for forest and farm producers (Macqueen and Korhalliler, 2011).

A few further examples are again illustrative of the problems faced. In Malawi, despite being the country's third largest industry by value (US$41.3 million annually) and employing 133,000 people, no charcoal licence has yet been issued by government. Roughly 12% of the final sales price to poor urban consumers consisted of informal payments along the value chain to government officials – and forest and farm producers received just 21% of the final sales price – none of which was reinvested in forest management, as charcoal production is deemed illegal (Kambewa *et al.*, 2007). A similar situation is found in Tanzania, where the total annual revenue generated by the charcoal sector is estimated at US$650 million to the country as a whole, dwarfing the contribution of coffee and tea to the national economy (at US $60 million and US$45 million respectively) (Peter and Sander, 2009). The scale of unregistered or unregulated activities in charcoal production and use lead to an estimated loss to the treasury of about US$100 million per year. In Senegal also, almost all households depend on charcoal. In 1998 the 170 presidents and treasurers of the market's 85 main charcoal cooperatives took most of the government's charcoal quotas. The 20 wealthiest merchants and

25 wealthiest wholesalers made on average US$60,000 and US$30,000 per year respectively, in comparison with an average forest charcoal producer who received on average US$41 – and then only if village chiefs distribute charcoal revenues fairly (Ribot, 1998). Even with decentralisation nominally devolving control over charcoal production to local rural councils, those monopolising market power have found ways to maintain their grip on the revenues (Ribot, 2009). Opening up market access through a process of equitable legalisation is clearly a critical governance challenge.

For locally controlled forestry enterprises wishing to access more lucrative export markets, buyer concerns over sustainability have often led to requirements for product certification. In some cases this is for organic certification of forest-based food crops (e.g. coffee, honey, brazil nuts, forest berries, ginseng, maple syrup and palm hearts). In other cases this is for sustainable timber production. Two main schemes emerged to certify sustainable forest management: from 1993, the Forest Stewardship Council (FSC), and from 1999, the Programme for the Endorsement of Forest Certification (PEFC). High per-unit costs of certification have plagued the application of certification to small and medium forest enterprises that make up the vast majority of enterprises within the broader ambit of locally controlled forestry. Various approaches to group certification have been one response to try and reduce these costs (Barr *et al.*, 2012).

In 2006, calls emerged for an alternative, trust-based, Fairtrade timber model in which buyers would cover certification costs for small and medium forest enterprises (Macqueen *et al.*, 2006a). The justification was that forest proximity, lack of mobility and social licence to operate in locally controlled forestry improves incentives for sustainability when compared with large-scale industrial forestry. It therefore requires a different certification model. Despite demand for such a model (see Macqueen *et al.*, 2008) fears over a backlash from the environmental forces behind mainstream forest certification led to a joint FSC–Fairtrade labelling scheme – in which neither the unique features of locally controlled forestry nor its cost disadvantages in certification were truly addressed.

Notwithstanding these constraints, great progress has been made by locally controlled forestry enterprises in accessing markets (Kozak, 2007). Yet, even for products that are allowed fair market access there is often a need to strengthen sustainable business capacity among forest and farm producers. In many developing country contexts there are few formal programmes and institutions to support locally controlled forest enterprises. Recent and ongoing attempts to network those institutions, build shared knowledge of how best to support locally controlled forest enterprises, and spread toolkits of best practice, require further resourcing (Inglis, 2013).

Good guidance now exists on how best to support locally controlled forest enterprises to access markets – in areas such as developing market understanding, participatory value chain analysis, product development, business planning and service provision, financial planning and service provision, strengthening enterprise organisation, building in ecological sustainability, and policy research to improve the enabling environment (Macqueen *et al.*, 2012b). The latter reference links to a vast array of manuals and toolkits for strengthening different aspects of locally controlled forestry business – for example the excellent group enterprise resource book (Bonitatibus and Cook, 1995) or the detailed market analysis and development guide (Lecup and Nicholsen, 2000).

Enabling condition 3 – Good quality technical extension services

Access to support services, especially extension and other types of technical capacity building is also essential. Forest and farm producers need to have confidence that their forest management efforts, and the development of processing, packaging and marketing techniques are being carried out in an optimal way, using the best available information, skills and technologies.

Technical extension services in forestry are carried out by foresters from a mix of government agencies, higher education institutions, private sector companies, Non-Government Organisations (NGOs) and forest producers' own organisations. In some countries extensionists are able to exchange experiences through professional societies set up for that purpose – but representation from private sector companies and NGOs is often limited.

In the 1980s, at the onset of the move towards 'social forestry' (a precursor to 'locally controlled forestry') it was acknowledged that traditionally, interactions between foresters and rural populations had been limited to protection, policing and revenue collection (Falconer, 1987). Where technical advice had been given it was mainly oriented towards tree planting, forest management and primary processing. At that time it was noted that the art of encouraging people's participation in forestry was new to foresters. The idea of group enterprises around a wide range of forest product subsectors was still in its infancy.

Forest extension practice has come a long way in the intervening period. More than a decade ago, the paternalistic view of extensionists as the external advisors or experts who instructed or directed the recipients or learners in what to do had largely been replaced by a model based on principles from the broad field of adult education: the learner has significant control throughout the entire learning process, based on the growing realisation that in an area such as improved forest management skills, much of what adult learners need to have they already know (Seevers *et al.*, 1997). An international survey of global forest extensionists noted four key strategies were widely regarded by forest extensionists as pivotal to their ability to interact positively with learners: involving the learner in planning, building trust, establishing rapport, and learning about the culture of the learners (Johnson *et al.*, 2007). But in terms of professional development, responses from tropical countries, while indicating some in-service training, showed much fewer opportunities for training leading to professional licensing or membership of professional societies. The latter area is important, because the exchange between institutions for higher education, private sector companies and NGOs can greatly increase the reach of best practice.

In terms of content, increasing guidance has emerged over time for the development of small-holder and community forestry extension for a range of government and civil society audiences in temperate and tropical regions. These guidance materials include: participatory approaches to engaging forest and farm producers; models of organisation management; guides to business development; and information on technical issues (including tree selection, managing forestry as common property and how to achieve certified sustainable forest management) (Jackson and Ingles, 1998; Arnold, 1998; Warner, 1999; Irwin, 2007; Communities Committee, 2008; Barr *et al.*, 2012; Mulkey and Day, 2012; Bobb-Prescott and Kumar, 2013). A full listing of extension manuals relevant to locally controlled forestry is beyond the scope of this chapter.

One of the biggest challenges in ensuring availability of more and better extension services to meet the needs of an expanding locally controlled forestry sector is that they are expensive. Especially during periods of restraint in government budgets, extension is often high on the list of expendable items. And yet, it is critically important in increasing the effectiveness of both other enabling investments and asset investments. One direction of promise is the development of partnership arrangements between government agencies and forest producer organisations.

Enabling condition 4 – Effective forest and farm producer organisations

This fourth enabling condition has a somewhat different character from the first three. Forest and farm producer organisations are a tool, normally created by producers themselves to achieve goals they cannot achieve as well or at all as individuals, through various kinds of economies of scale. These form the basic purposes or functions of any local forest and farm producer organisation (Macqueen *et al.*, 2006b; deMarsh *et al.*, 2014) and may in a mature organisation also constitute the main pillars of its revenue generation:

- *To speak with a single more powerful voice/lobby for more supportive policies:* especially those relating to enabling conditions 1–3: broadened and strengthened tenure, better market access, and improved extension.
- *To reduce transaction costs and provide services for their members:* forest protection, forest management, various marketing functions (negotiating sales, aggregating product, improving product quality, storage and transportation), forest extension education and training, and other types of capacity building.
- *To adapt strategically to new opportunities through value added processing of forest products.*

These broad purposes or functions will be pretty much the same for a producer of timber from a family forest in Norway and a producer of wild forest coffee or honey from a community forest in Ethiopia. The first function is often the initial rationale for forest and farm producer organisations, and the basis for revenues organised through membership fees (or tariffs on products sold through that organisation). Lobbying governments to put the first three enabling conditions in place, and over time, to ensure they stay in place, and adapt to changing circumstances has an 'open-ended' quality. Priorities may change over time; new challenges will likely emerge that were not anticipated in the beginning but in response to which, the lobbying capacity of the forest and farm producer organisation may be as important as for its early goals. The second function may often relate to services that the organisation then arranges, initially on a 'not-for-profit' basis, but which may later become a source of revenue, potentially sold to customers outside the organisation. The third function relates to the enterprises associated with processing forest products and will likely include a profit-making objective (that is often part of a more or less formal 'triple bottom line').

Forest and farm producer organisations share the need for similar enabling conditions so it is not surprising to find a few common types with some frequency (FAO, 2013):

- Informal village-level forest management labour-sharing groups
- Formally constituted village-level cooperatives for improved forest management capacity or marketing into local markets
- County/district/township-level associations for marketing into regional and national markets; and forest management support services such as forest extension
- Regional/national federations of local-level forest and farm producer organisations that serve a lobbying function

Forest and farm producer organisations are not an end in themselves. A lot of time and other resources are needed to establish and maintain them. Producers know this, either from previous historical experience, or from observing the experience of other groups, and know the challenges of maintaining or failing to maintain good governance within the organisation, and of reducing conflict within the group of producers, and with other groups, communities, the private sector and the government. An *effective* forest and farm producer organisation, therefore has the ability to get results in a way that at the very least does not increase the various kinds of conflict that may be present in their context. To form a forest and farm producer organisation, producers must really believe it is worth the effort and the risks.

The advantages and challenges of producer organisations in the forest sector are now well known, with a wealth of helpful case studies documented from around the world (see FAO, 2013). Inevitably, forest and farm producer organisations vary in character in a number of ways that reflect national and local history and culture, and the specific activities and products that are the focus of their work. These important differences co-exist with other qualities that seem to be quite universal in their association with the effectiveness of forest and farm producer organisations over time (deMarsh *et al.*, 2014):

- *Strong collective interests* which may evolve over time
- *Autonomous functioning* in relation to government and other agencies and institutions
- *Democratic decision-making* (one member, one vote; regular opportunities for members to discuss and approve or change basic organisation policies and strategy)
- *Transparent financial reporting* by organisational management to members
- *Successful experiences* from the past or present across members
- *Self-reliance in financial needs* for basic organisational functioning
- *Significance in the total economy* in which they operate
- *Subsidiarity* in the functions performed by the organisation: only perform functions at a secondary level (such as a federation of local forest and farm producer organisations) that cannot be done as well at a level closer to the producers (by the local forest and farm producer organisations themselves)

The geographic scale at which a forest and farm producer organisation – or federation – seeks to operate is determined again by the three basic functions: the level of government whose policy the forest producers would like to influence, the size of area that would allow a particular service to be provided most efficiently, and the area required to supply any demand from value added processing facilities. The development of national federations of local organisations has evolved at a fairly early stage in most cases of long-term success of locally controlled forestry (Nepal, Federation of Community Forestry Users in Nepal – FECOFUN; Finland, Maa- ja metsätaloustuottajain Keskusliitto – MTK; Sweden, Lantbrukarnas Riksförbund – LRF).

The challenge of maintaining a spirit of ownership and a sense of control by the producers themselves over time can easily increase with the size of the organisation. It increases even more dramatically as additional organisational layers are added, such as a provincial federation of local organisations and a national federation of provincial groups. Beyond the village level, organisations normally adopt a formal legal structure as cooperatives, business or not-for-profit associations or corporations, according to the decision by members as to the form that best suits their needs, goals and capacities, and the options available in the laws of their province or country. But while there are a vast array of informal groups of various sizes and functions that support family and community needs in critically important ways, there are also a growing number of formal organisations (18,000 community forest user groups in Nepal, 115,000 forest farmer cooperatives in China).

A final consideration in the effectiveness of a forest and farm producer organisation is the composition of the membership (deMarsh *et al.*, 2014). This must balance the call for uniformity of interests (to ensure goals are narrow enough, and cohesion strong enough to achieve success), with calls for breadth, e.g. both larger and smaller producers (to provide greater strength).

What can be done to encourage investing in locally controlled forestry?

An important starting point is the recognition that the argument for the advantages of locally controlled forestry is by no means won (despite substantial recent evidence in its favour from a variety of family, community and indigenous contexts – some of which has been highlighted above). There is therefore a pressing need to counter negative perceptions of the transaction costs and complexity of locally controlled forestry (in comparison with large-scale industrial forestry) with further evidence in support of three core arguments.

First, in terms of economic consumption and production – locally controlled forestry offers a model that delivers greater economic equity than large-scale industrial forestry. It does this in part because it involves local producers and rights-holders as owners, not just labourers. This is important because compelling evidence points to the current differential between rich and poor being provably bad for society in terms of infant death rates, child well-being, high school drop-out rates, social mobility between classes, development aid per capita, trust, mental illness, obesity, incidence of drug use, homicide, criminal imprisonment rates, and so on (Wilkinson and Pickett, 2010). In the forest sector, inequality is also linked to forest-related conflicts (de Koning *et al.*, 2008). Moreover, there is strong evidence that material wealth doesn't make us happier above a relatively low threshold (Clark *et al.*, 2008; Pretty, 2013).

The paradigm shift involved in ILCF is therefore important, not only because it is highly productive in generating economic wealth, but also because it distributes that economic wealth more equitably – which is good for rich and poor alike.

Second, in relation to social justice in the governance of public goods – locally controlled forestry delivers strong membership-based organisations within multi-functional landscape mosaics. Such organisations can represent the interests of local people in decision-making processes and in defining legality. This can help protect local public goods, and improve prospects for the formalisation of enterprises supplying them, in the face of growing competition from those pursuing global public goods (e.g. the supply of export crops or the conservation of the global climate). This too is important, because there are growing land and resource grabs to meet the needs of the already affluent (Cotula, 2013). Such land and resource grabs disrupt the social fabric of local communities, often dispossessing them altogether. In contrast, locally controlled forestry delivers a multiplicity of products and services through local businesses – that accrue and reinvest capital locally – fostering the important social fabric and interconnections that maintain rural communities (Macqueen *et al.*, 2014a).

Third, in relation to climate change adaptation and mitigation (and the functionality of critical ecosystem services), locally controlled forestry delivers a range of income-generating options for forest and farm producers. This diversity improves options for climate change adaptation and creates incentives to restore and maintain functional ecosystems on which they are based. And as noted above, locally controlled forestry is at least as good as alternative models for maintaining forest contributions to global climate change mitigation (Porter-Bolland *et al.*, 2012; Seymour *et al.*, 2014). This matters, because increasing population densities and consumption patterns will require increasingly intense production of a diverse range of products and services from finite landscapes – so aligning economic development with local incentives for forest protection is fairly fundamental. Additionally, any payment mechanisms for Reducing Emissions for Deforestation and forest Degradation (REDD+) will need to engage strong forest and farm producer organisations to halt deforestation and restore forest condition – and these are pursued as one of the four key areas of locally controlled forestry. In contrast, large-scale industrial forestry is often associated with increased economic inequality, reduced local control of public goods, and reduced income options for forest and farm producers. In situations where large industries provide forest and farm producers with important markets for forest-based raw materials, they also have significant market power to keep both wages and the prices they pay for these products low.

Clearly, investing in locally controlled forestry is not a peripheral issue. Rather, it is the central modality through which business might become sustainable – for forests and beyond – not just environmentally but social and economically. The current twofold approach, whereby large-scale industrial forestry is treated as the norm, and locally controlled forestry as a peripheral irrelevance, requires urgent reassessment. Overseas Development Aid needs to reassess whether the future is best brought about by constraining what is bad in the former model – or by building on what is good in the latter. A notable issue is that large-scale industrial forestry is often run by those with easy access to provincial, national and international decision-makers. For example, in Rio+20 where those controlling development finance discussed different approaches to forests and the green economy, it was notable how well-resourced large-scale forest industries dominated discussion platforms. So what is important is not just the capacity to make the conceptual arguments outlined above, but also to resource those who might represent those arguments in decision-making fora.

There is much that can be practically done to establish the four integrated enabling conditions for locally controlled forestry. Perhaps the best place to start – given the integrated logic outlined above, is to commit resources to strengthening and involving in decision-making the forest and farm producer organisations – which are generally established precisely to secure the other three enabling conditions. Such support might involve a range of different investments – which will depend very much on the stage of institutional development of the producer organisation in question (see Macqueen *et al.*, 2014b). Indeed a number of different actors can engage to support strong producer organisations: the producer organisations themselves, service providers and government (see FAO, 2014).

Forest and farm producer organisations can develop their own capacity through: democratic governance; learning from the experiences of farmer organisations; avoiding overreliance on external funding; demonstrating tangible benefits for members; formulating clear, solution-oriented messages for policy-makers; collaborating and building trust with diverse partners (including governments); and creating federations or seeking better ways to coordinate with other producer organisations.

Service providers, which may include producer organisations, NGOs, consultants, academics, private companies, government agencies and development partners, can help create an enabling environment by: supporting the creation of producer organisations; dealing fairly with them; offering a wide variety of useful training; facilitating peer–peer exchanges; compiling and spreading useful guidance on best practices; partnering with producer organisations financially or technically; and facilitating connections with buyers, lenders and other service providers.

Governments can help producer organisations by creating an enabling environment by: ensuring producers have secure and long-term access and tenure to forests, land and trees; removing legislative and regulatory barriers; giving guidance and the highest possible degree of freedom at the local level for the establishment of producer organisations; respecting their contribution to policy processes and their right to federate; establishing dedicated grant mechanisms and service provision; ensuring fair law enforcement; preferentially favouring producer organisations in government purchasing policies; and generally matching policies with the interests of producer organisations.

Enabling investment to develop strong forest and farm producer organisations can be catalytic in helping put in place the other three enabling conditions described above. When the four enabling conditions are in process of being firmly established in the locally controlled forestry sector, self-sustaining transformation begins. This has occurred in the Nordic countries in the first half of the past century, and more recently in Nepal, with Guatemala and China showing strong indications of entering this phase as well. Many more countries, such as the Gambia, Mexico and Tanzania are also heading in this direction.

The four enabling conditions contribute to improvements in livelihoods and forest protection simultaneously, and in a way which is mutually reinforcing. Producer energies are released, and it becomes much easier to attract other requirements for development, such as asset investment and private sector partnerships to further add value to locally controlled forestry and the businesses derived from it. For a practical guide for investors on necessary steps for matching enabling investment with asset investment see Elson (2013). Aligning business (and development more broadly) with sustainable management of the world's forests is indeed a pressing challenge – and one that locally controlled forestry can rise to meet with the right mix of enabling investment.

Note

1 Our definition – 'forest and farm producers are members of families, communities or indigenous peoples with rights to forests on which they more or less regularly depend for some or all of their subsistence and income needs'. In much of the world, these forest producers are also farmers whose forest-based production is more or less integrated into a single production unit or enterprise.

References

Ackzell, L. (2009) '100 years of Swedish forest owner associations, challenges ahead'. Presentation to the CIFOR international conference 'Taking stock of smallholder and community forestry: Where do we go from here?' 24–26 March 2010, Montpellier, France, Federation of Swedish Family Forest Owners, Stockholm, Sweden.

ANSAB (2009) *Challenges and opportunities for Nepal's small and medium forest enterprises (SMFEs)*. Asia Network for Sustainable Agriculture and Bioresources, Kathmandu, Nepal.

Arnold, J.E.M. (1998) Managing forests as common property. FAO forestry Paper 136. FAO, Rome, Italy.

Barr, R., Busche, A., Pescott, M., Wiyono, A., Putera, A.E., Victor, A., Novi Fauzan, B., Prantio, S. and Karnanto, U. (2012) *Sustainable community forest management – A practical guide to FSC group certification for smallholder agroforests*. University Book Store Press, Seattle, USA.

Bobb-Prescott, N. and Kumar, H. (2013) *Participating in managing forests – A guide to community forestry in the Caribbean islands*. Caribbean Natural Resources Institute (CANARI) Guidelines series, Laventille, Caribbean.

Bonitatibus, E. and Cook, J.F. (1995) *The group enterprise resource book*. FAO, Rome, Italy.

Bowler, D., Buyung-Ali, L., Healey, J.R., Jones, J.P.G., Knight, T. and Pullin, A.S. (2010) *The evidence base for community forest management as a mechanism for supplying global environmental benefits and improving local welfare: Systematic review*. CEE review 08-011 (SR48). Environmental Evidence. Available at: www.environmentalevidence.org/SR48.html. Accessed 8 May 2014.

Bray, D., Negreros, P., Merino-Perez, L., Torres Rojo, J.M., Segura, G. and Vester, H.F.M. (2003) Mexico's community managed forests as a global model for sustainable forestry. *Conservation Biology* 17(3): 672–677.

Bulkan, J. (2014) Forest grabbing through forest concession practices: the case of Guyana. *Journal of Sustainable Forestry* 33(4): 407–434.

CEESP (2008) *Recognising and supporting indigenous and community conservation – ideas and experiences from the grassroots*, CEESP Briefing Note 9. Available at: http://www.rightsandresources.org/documents/files/doc_1049.pdf.

Chao, S. (2012) *Forest peoples – numbers across the world*. Forest Peoples Programme, Moreton-in-Marsh, UK.

Charnley, S. and Poe, M. (2007) Community forestry in theory and in practice: where are we now? *Annual Review of Anthropology* 26: 301–336.

Clark, A., Frijters, P. and Shields, M. (2008) Relative income, happiness, and utility: an explanation for the Easterlin paradox and other puzzles. *Journal of Economic Literature* 46(1): 95–144.

Communities Committee (2008) *Acquiring and managing a community-owned forest: A manual for communities*. Communities Committee, Baltimore, USA.

Cotula, L. (2013) *The great African land grab? Agricultural investment and the global food system*. Zed Books, London, UK.

de Koning, R., Capistrano, D., Yasmi, Y. and Cerutti, P. (2008) *Forest-related conflict – Impact, links, and measures to mitigate*. Rights and Resources Initiative, Washington, D.C., USA.

deMarsh, P., Boscolo, M., Savenije, H., Campbell, J., Zapata, J., Grouwels, S. and Macqueen, D. (2014) Making change happen – how governments can strengthen forest producer organisations. Forest and Farm Facility Working Paper, FAO, Rome, Italy.

Elson, D. (2013) *Guide to investing in locally controlled forestry*. Growing Forest Partnerships, London, UK.

Falconer, J. (1987) Forestry extension: a review of the key issues. Overseas Development Institute Network Paper 4e. Overseas Development Institute, London, UK.

FAO (2012) *Voluntary guidelines on the responsible governance of tenure of land fisheries and forests in the context of national food security*. Food and Agriculture Organization of the United Nations, Rome, Italy.

FAO (2013) *Strength in numbers: Effective forest producer organizations*. Food and Agriculture Organization of the United Nations, Rome, Italy.

FAO (2014) Strength in numbers. Summary statement from an international conference on forest producer organizations held in Guilin, Guangxi Autonomous Region, China, 25–28 November 2013. Food and Agriculture Organization of the United Nations, Rome, Italy. Available at: www.fao.org/partnerships/forest-farm-facility/39494-0b7c58dc033ba4d8f99d595cd33b1bac7.pdf. Accessed 14 May 2014.

G3 (2011) Memorandum of Understanding between the Global Alliance of Community Forestry, GACF, the International Alliance of Indigenous and Tribal Peoples of the Tropical Forest, IAITPTF, and the International Alliance of Family Forestry, IFFA. The Three Rights Holders' Group – G3, Lassnitzhöhe, Austria. Available at: www.g3forest.org/userfiles/file/G3/G3PlanningMeetingOct2011/G3MOUD2.pdf. Accessed 7 May 2014.

Gebremariam, A.H., Bekele, M. and Ridgewell, A. (2009) Small and medium forest enterprises in Ethiopia. IIED Small and Medium Forest Enterprise Series No. 26. FARM-Africa and IIED, London, UK.

Gilmour, D. and Fisher, B. (2011) *Reforming forest tenure: Issues, principles and process*. FAO Forestry Paper 165. Food and Agriculture Organization of the United Nations, Rome, Italy.

Inglis, A. (2013) *An independent review of Forest Connect*. ASI Training and Consultancy, Edinburgh, UK.

Irwin, B. (2007) *The key steps in establishing participatory forest management: A field manual to guide practitioners in Ethiopia*. Farm-Africa and SOS Sahel, Addis Ababa, Ethiopia.

Jackson, W. and Ingles, A. (1998) *Participatory techniques for community forestry: A field manual*. IUCN, Gland, Switzerland.

Johnson, J.E., Creighton, J.H. and Norland, E.R. (2007) An international perspective on successful strategies in forestry extension: A focus on extensionists. *Journal of Extension* 45(2). Available at: www.joe.org/joe/2007april/index.php. Accessed 8 May 2014.

Kambewa, P., Mataya, B., Sichinga, K. and Johnson, T. (2007) *Charcoal the reality – A study of charcoal consumption, trade and production in Malawi*. IIED, London, UK.

Kozak, R. (2007) *Small and medium forest enterprises: Instruments of change in a developing world*. Rights and Resources Initiative, Washington, D.C., USA.

Larson, A. and Dahal, G. (2013) Forest tenure reform: new resource rights for forest-based communities? *Conservation and Society* 10(2): 77–90.

Lawry, S., McLain, R., Swallow, B. and Biedenweg, K. (2011) *Devolution of Forest Rights and Sustainable Forest Management, Vol. 1: A review of policies and programs in 16 developing countries*. United States Agency for International Development (USAID), Tetra Tech ARD, Burlington, USA.

Lecup, I. and Nicholsen, K. (2000) *Community-based tree and forest product enterprises: Market analysis and development (MA&D) field manual. Books A–F*. FAO, Rome, Italy.

Long, C. (2010) *Forests and community control: Official processes that permit formal recognition of community management and rights, and their relevance to the Democratic Republic of Congo*. IIED and Forests Monitor, Cambridge, UK.

Macqueen, D. (2007) *The role of small and medium forest enterprise associations in reducing poverty*. In 'A cut for the poor' – Proceedings of the international conference on managing forests for poverty reduction: Capturing opportunities in forest harvesting and wood processing for the benefit of the poor. Ho Chi Minh City, Viet Nam, 3–6 October 2006.

Macqueen, D. (2013a) Landscapes for public goods: Multifunctional mosaics are fairer by far. IIED Briefing. IIED, London, UK.

Macqueen, D. (2013b) Enabling conditions for successful community forestry enterprises. *Small-scale forestry* 12(1): 145–163.

Macqueen, D.J. and Korhalliler, S. (2011) Bundles of energy: The case for renewable biomass energy. NR series No. 24. IIED, London, UK.

Macqueen, D., Dufey, A. and Patel, B. (2006a) Exploring fair trade timber: A review of issues in current practice, institutional structures and ways forward. IIED Small and Medium Forestry Enterprise Series No. 19. IIED, Edinburgh, UK.

Macqueen, D., Bose, S., Bukula, S., Kazoora, C., Ousman, S., Porro, N. and Weyerhaeuser, H. (2006b) Working together: Forest-linked small and medium enterprise associations and collective action. IIED Gatekeeper Series No. 125. IIED, London, UK.

Macqueen, D., Dufey, A., Gomes, A.P.C., Nouer, M.R., Suárez, L.A.A., Subendranathan, V., Trujillo, Z.H.G., Vermeulen, S., Voivodic, M. de A. and Wilson, E. (2008) Distinguishing community forest products in the market: Industrial demand for a mechanism that brings together forest certification and fair trade. IIED Small and Medium Forestry Enterprise Series No. 22. IIED, Edinburgh, UK.

Macqueen, D., Buss, C. and Sarroca, T. (2012a) *TFD Review: Investing in locally controlled forestry*. The Forest Dialogue, New Haven, USA.

Macqueen, D.J., Baral, S., Chakrabarti, L., Dangal, S., du Plessis, P., Griffiths, A., Grouwels, S., Gyawali, S., Heney, J., Hewitt, D., Kamara, Y., Katwal, P., Magotra, R., Pandey, S.S., Panta, N., Subedi, B. and Vermeulen, S. (2012b) Supporting small forest enterprises – A facilitators toolkit. Pocket guidance not rocket science! IIED Small and Medium Forest Enterprise Series No 29. IIED. London, UK.

Macqueen, D., Andaya, E., Begaa, S., Bringa, M., Greijmans, M., Hill, T., Humphries, S., Kabore, B., Ledecq, T., Lissendja, T., Maindo, A., Maling, A., McGrath, D., Milledge, S., Pinto, F., Quang Tan, N.,Tangem, E., Schons, S. and Subedi, B. (2014a) *Prioritising support for locally controlled forest enterprises*. IIED, London, UK.

Macqueen, D., Campbell, J. and deMarsh, P. (2014b) The forest and farm facility: Building strength in numbers. IIED Briefing, IIED, London, UK.

Mayers, J. (2006a) Poverty reduction through commercial forestry: What evidence? What prospects? Tropical Forest Dialogue Background Paper. The Forest Dialogue, New Haven, USA.

Mayers, J. (2006b) Small and medium-sized forest enterprises. Are they the best bet for reducing poverty and sustaining forests ? *International Tropical Timber Organisation Tropical Forest Update* 16(2): 10–11.

Mayers, J., Morrison, E., Rolington, L., Studd, K. and Turrall, S. (2013) Improving governance of forest tenure: a practical guide. Governance of Tenure Technical Guide No.2, IIED, and Food and Agriculture Organization of the United Nations, London and Rome.

Molnar, A., Liddle, M., Bracer, C., Khare, A., White, A. and Bull, J. (2007) Community-based forest enterprises: their status and potential in tropical countries. ITTO Technical Series No. 28, International Tropical Timber Organization, Yokohama, Japan.

Mulkey, S. and Day, J.K. (2012) *The community forestry guidebook II: Effective governance and forest management*. FORREX Forum for Research and Extension in Natural Resources, Kamloops, Canada.

Nepstad, D., Schwartzman, S., Bamberger, B., Santilli, M., Rar, D., Schlesinger, D., Lefebvre, P., Alencar, A., Prinz, E., Fiske, G. and Rolla, A. (2006) Inhibition of Amazon deforestation and fire by parks and indigenous lands. *Conservation Biology* 20(1): 65–73.

Nhancale, B.A., Mananze, S.E., Dista, N.F., Nhantumbo, I. and Macqueen, D.J. (2009) Small and medium forest enterprises in Mozambique. IIED Small and Medium Forest Enterprise Series No. 24. Centro Terra Viva and IIED, London, UK.

Ojha, H.R., Pandey, G., Dhungana, S., Silpakar, S. and Sharma, N. (2009) Investing in community managed forestry for poverty reduction in Nepal: A scoping of investment opportunities. Paper presented at the second international dialogue on Investing in Locally Controlled Forestry, 21–24 September 2009, Kathmandu, Nepal. Available at: http://tfd.yale.edu/sites/default/files/tfd_ilcf_nepal_background paper.pdf. Accessed 7 May 2014.

Openshaw, K. (2010) Can biomass power development? Gatekeeper 144: April 2010. IIED, London, UK.

Osei-Tutu, P., Nketiah, K., Kyereh, B., Owusu-Ansah, M. and Faniyan, J. (2010) Hidden forestry revealed: Characteristics, constraints and opportunities for small and medium forest enterprises in Ghana. IIED Small and Medium Forest Enterprise Series No. 27. Tropenbos International and IIED, London, UK.

Pagdee, A., Kim, Y. and Daugherty, P.J. (2006) What makes community forest management successful? A meta-study from community forests throughout the world. *Society and Natural Resources: An International Journal* 19(1): 33–52.

Persha, L., Agrawal, A. and Chhatre, A. (2011) Social and ecological synergy: local rulemaking, forest livelihoods, and biodiversity conservation. *Science* 331(6024): 1606–1608.

Peter, C. and Sander, K. (2009) *Environmental crisis or sustainable development opportunity? Transforming the charcoal sector in Tanzania*. World Bank, Washington, D.C., USA.

Porter-Bolland, L., Ellis, E.A., Guariguata, M.R., Ruiz-Mallen, I., Negrete-Yanelevich, S. and Reyes-Garcia, V. (2012) Community managed forests and forest protected areas: an assessment of their conservation effectiveness across the tropics. *Forest Ecology and Management* 268: 6–17.

Pretty, J. (2013) The consumption of a finite planet: well-being, convergence, divergence and the nascent green economy. *Environmental Resources Economics* 55: 475–499.

Raworth, K. (2012) A safe and just space for humanity. Oxfam Discussion Papers. Oxfam, Oxford, UK.

Ribot, J. (1998) Theorizing access: forest profits along Senegal's charcoal commodity chain. *Development and Change* 29: 307–341.

Ribot, J. (2009) Authority over forests: empowerment and subordination in Senegal's democratic decentralization. *Development and Change* 40(1): 105–129.

Robinson, B., Holland, M. and Naughton-Treves, L. (2011) *Does secure land tenure save forests? A review of the relationship between land tenure and tropical deforestation*. CCAFS Working Paper No. 7. CGIAR Research Program on Climate Change, Agriculture and Food Security (CCAFS), Copenhagen, Denmark.

Rockström, J., Steffen, W., Noone, K., Persson, Å., Chapin, III, F.S., Lambin, E., Lenton, T.M., Scheffer, M., Folke, C., Schellnhuber, H., Nykvist, B., De Wit, C.A., Hughes, T., van der Leeuw, S., Rodhe, H., Sörlin, S., Snyder, P.K., Costanza, R., Svedin, U., Falkenmark, M., Karlberg, L., Corell, R.W., Fabry, W.J., Hansen, J., Walker, B., Liverman, D., Richardson, K., Crutzen, P. and Foley, J. (2009) Planetary boundaries: exploring the safe operating space for humanity. *Ecology and Society* 14(2): 32.

RRI (2012) *What rights? A comparative analysis of developing countries' national legislation on community and indigenous peoples' forest tenure rights*. Rights and Resources Initiative, Washington, D.C., USA.

RRI (2014) *What future for reform? Progress and slowdown in forest tenure reform since 2002*. Rights and Resources Initiative, Washington, D.C., USA.

RRI and ITTO (2009) *Tropical forest tenure assessment: Trends, challenges and opportunities*. Rights and Resources Initiative, Washington, D.C., USA, and International Tropical Timber Organization (IITO), Yokohama, Japan.

Scherr, S., White, A. and Kaimowitz, D. (2003) Making markets work for forest communities. *International Forestry Review* 5(1) 67–73.

Seevers, B., Graham, D., Gamon, J. and Conklin, N. (1997) *Education through cooperative extension*. Delmar Publishers, Albany, New York.

Seymour, F., La Vina, T. and Hite, K. (2014) *Evidence linking community level tenure and forest condition: An annotated bibliography*. Climate and Land Use Alliance, Washington, D.C., USA.

Tint, K., Springate-Baginski, O., Macqueen, D.J. and Gyi, M.K.K. (2014) *Unleashing the potential of community forest enterprises in Myanmar*. Ecosystem Conservation and

Community Development Initiative (ECCDI), University of East Anglia and IIED, London, UK.

Warner, K. (1999) *Selecting tree species on the basis of community needs.* FAO Community Forestry Field Manual 5. FAO, Rome, Italy.

Wilkinson, R. and Pickett, K. (2010) *The spirit level: Why equality is better for everyone.* Penguin Books, London, UK.

8 Decentralization and community-based approaches

Reem Hajjar and Augusta Molnar

Decentralization of forest management has been a major trend in global forest governance for the past three decades (Agrawal *et al.* 2008; Ribot *et al.* 2006). Decentralization is the process by which a central government cedes powers to actors and institutions at lower levels of government (Mawhood 1983, cited in Ribot *et al.* 2006). Political or democratic decentralization entails the transfer of power to actors or institutions that are politically accountable to the population in their jurisdiction (Ribot *et al.* 2006; Sharma 2006). Decentralization changes existing power structures (Raik *et al.* 2008), and can be expected to empower citizens, especially disadvantaged groups, in relation to a distant government (Samoff 1990). In this context, this chapter addresses the power shift from central governments to communities in managing forests that they live in and around; we look at community-based forest management as a form of democratic decentralization of forest governance – the transfer of power over forest resources and management to local governments and authorities more representative of local populations.

National governments have sought to decentralize power over many services, including management of vast forest resources that provide substantial national income in many regions. Reasons for this decentralization include: appeasing demands from international donors, NGOs and local citizens demanding better governance by enhancing public sector transparency and accountability; reducing costs of overextended central bureaucracies; succumbing to pressure to right the wrongs of commercial forestry that excluded local people from accessing and benefiting from forest resources; and acknowledging the potential benefits to conservation and rural development that more participatory approaches to resource management can provide (Agrawal and Ribot 1999; Brown *et al.* 2010; Manor 1999). Decentralization has been seen as a way to value and legitimize traditional ecological knowledge in modern natural resource management (Charnley and Poe 2007).

As a result, more than three-quarters of developing countries and countries in transition are currently experimenting with decentralization of natural resource management (Contreras-Hermosilla *et al.* 2006; Ribot 2004). As of 2013, over 513 million hectares of forests (or 15.5% of all forests) were held globally by Indigenous Peoples and local communities under some form of statutory[1] community ownership or control. In the lower- and middle-income countries (LMICs) the figure is higher: 30.1% of official forest lands (Rights and Resources Initiative 2014). Areas under community or Indigenous Peoples conservation and management – whether informal or recognized by the state – are greater than public protected areas in Africa, Asia and Latin America (Molnar *et al.* 2004).

These recent trends in decentralization of natural resource management have created the context in which community forestry has been able to proliferate. Community forestry, or community-based forest management, is "the sustainable management of forests for wood, non-timber forest products and other social or environmental service values, carried out by forest-dependent families or smallholders, community groups and indigenous peoples" (Growing Forests Partnership n.d.). With formal community forestry, "some degree of responsibility and authority for forest management is formally vested by the government in local communities", and a central objective is to provide local social and economic benefits derived from forests (Charnley and Poe 2007). Community forestry is a form of community-based natural resource management (CBNRM), which commonly refers to the management and conservation of natural resources being devolved to local and/or indigenous institutions and people, while linking socio-economic development with environmental conservation (Kellert *et al.* 2000).

Community-based forest management has been promoted as a model to create long-term economic development and self-reliance in rural communities, while promoting the conservation and sustainable use of forests and consolidating rights over lands and resources (Bray *et al.* 2008; Pagdee *et al.* 2006; Scherr *et al.* 2004). In many cases around the world, it has improved efficiency, equity, democracy and ecosystem health in forest-dependent communities (Larson 2005; Molnar *et al.* 2007), while in many other cases, community forestry has failed to achieve its goals.

In this chapter, we will explore some of the reasons why community forestry has emerged, how a shift from centralized ownership models to community-controlled models may help to achieve sustainability, and why some decentralized approaches are successful while others continue to struggle.

Why decentralize?

There are many reasons why this trend of decentralizing forest governance has emerged. In many instances, it has been in response to failures of centralized governments to effectively and efficiently manage state-owned forests. Many state governments have recognized that they are too weak or too resource poor to enforce laws and regulations across state-owned forests, and thus aim to decentralize responsibilities and costs to private entities, lower levels of government, or community groups (Charnley and Poe 2007). It has been argued that community forestry came about in British Columbia, a province of Canada, due to neoliberalist ideals, challenging the command-and-control regulatory approach to forest governance (McCarthy 2006). In The Gambia, decentralization was part of a deliberate financial strategy to decentralize costs of forest service delivery and forest management to local constituencies (Schroeder 1999). In Mexico, forest management decentralization was partly due to a structural adjustment plan in the 1970s and 1980s that required the dismantling of state-owned companies, including parastatal forestry companies that had been operating 25-year concessions on community-owned but state-managed forest land (Chapela 2005). In other instances, pressure to decentralize came from donors in response to perceived failures of the state: Cameroon's forestry sector underwent a series of liberal reforms in the early 1990s prescribed by the World Bank (Oyono 2004). Studies have shown that in many African countries, this prevailing model of state-owned forests, often managed under export-oriented industrial concession models, contributes little to widespread growth, development and rural poverty reduction (Molnar *et al.* 2010). However, this model continues to prevail in many countries, based on the idea that it will optimize the contribution of forests to overall national needs and interests through export-led economic growth.

In many countries, rather than having been pushed solely by donors or central governments, decentralization has been a result of grassroots movements and local and indigenous groups' struggles for the recognition of their traditional rights to land and resources or to protest destructive resource extraction practices (reviewed in Charnley and Poe 2007). This shift recognizes local and indigenous people not only as stakeholders in forest management but as rights-holders (Colchester 2004). Those advocating for decentralization of forest governance often consider the participation, political power and property rights of marginalized peoples an important objective (Kellert *et al.* 2000). Importantly, Colchester (2008) advocates a rights-based approach that goes beyond tenure itself. It is not adequate just to ensure property rights – the broader spectrum of rights needs to be considered, including protection of cultural, social, livelihood, civil and political rights and freedoms.

Particularly in Latin America, recognition of indigenous and local land rights has been a major reason for shifting forest tenure rights from states to communities in the last 20 years (Larson *et al.* 2008). Grassroots resistance to forest concession policy contributed to the demise of industrial concessions and the rise of community forestry in the 1970s and 1980s in parts of Mexico (Chapela 2005). The Amazonian rubber tappers' movement of the 1980s led to the establishment of extractive reserves in Brazil that guaranteed forest use and exclusion rights to non-indigenous groups who derive their livelihoods from forest use (Keck 1995). Indigenous Peoples are becoming stronger advocates for decentralized forest management regimes that recognize their tenure ownership and rights. Local resistance in the Petén, Guatemala, led to the allocation of community concessions in the Mayan Biosphere Reserve, from which the government had originally excluded them

(Barsimantov *et al.* 2011). In India, some communities were granted new usufruct rights following many years of conflict around the failure to recognize forest communities' tenure (Larson and Dahal 2012), although these rights were quite limited.

Community forestry for conservation and poverty alleviation: In theory …

These decentralization trends are the context in which community-based natural resource management (CBNRM), and community forestry in particular, has proliferated. In many countries where the state has failed to effectively manage forests, or where traditional rights to land and forest resources have been recognized, community forestry has been promoted as a better way both to manage forests, and to sustain rural development and alleviate poverty among rural populations. In many geographic locations, communities and Indigenous Peoples have long been the *de facto* managers of forest resources and have long argued that the continuation of forest cover in their territories is a testimony to their effective forest management and conservation; yet they often lack legitimacy in law or policy.

Theoretically, there are many reasons why community forestry and CBNRM are attractive concepts for conservation (reviewed in Bradshaw 2003; Charnley and Poe 2007; Kellert *et al.* 2000). In response to the perceived failings of centralized, or top-down management, it is argued that local communities could do a better job of managing local resources sustainably. Indeed, their proximity to the resources amplifies the repercussions of their decisions and creates a vested interest in maintaining their environment over the long term. It also enables quicker adaptation to natural resource fluctuations or changing circumstances. Locals can identify and prioritize pressing environmental problems (Larson 2003), and more (cost) effectively monitor resource use and enforce rules (reviewed in Crona *et al.* 2011). Oftentimes, local managers bring with them a local or traditional ecological knowledge that is particularly attuned to the particular environment through years or generations of experience (Berkes *et al.* 1994, cited in Kellert *et al.* 2000). Locals are more apt to abide by resource-use rules if they are the ones creating those rules, making management choices more locally legitimate (Larson 2003; Schlager and Ostrom 1992).

Furthermore, CBNRM has gained credibility as a way to promote socio-economic development in rural areas through conservation or natural resource management (reviewed in Charnley and Poe 2007; Kellert *et al.* 2000). A central tenet of community forestry and CBNRM is that local communities benefit socially and economically from the resource they manage, providing them with multiple incentives for good management.

Decentralized, participatory approaches to resource management, such as CBNRM, have been proposed as an effective way to deal with uncertainty, complexity and change. Including a broader set of stakeholders reflects the need for new modes of governance that can access different kinds of knowledge bases to deal with complex social–ecological systems (Berkes *et al.* 2003; Crona *et al.* 2011). Pahl-Wostl (2009) found that resource management regimes that include bottom-up governance processes with strong stakeholder participation tend to be characterized by higher levels of social learning and increased adaptive capacity. In dealing with a changing climate and changing political and market contexts, such decentralized approaches will likely play an increasingly important role in resource management.

However, this community-based approach has also been met with criticism. Not all communities prove to be good managers. Better management of the resource is predicated on a community's credibility, capacity and genuine interest in managing local resources in the interests of all stakeholders, including future generations and non-locals (Bradshaw 2003). In a decentralized context, communities may not manage a forest sustainably if underlying economic incentives for deforestation continue to preside (Tacconi 2007), or if local populations find globally relevant concerns such as biodiversity conservation or climate change mitigation to be of limited relevance to them (Ritchie, 1996, as cited in Bradshaw, 2003). Further, overly decentralized management may foster decisions that miss the "bigger picture" needed to manage landscapes, and fail to complement the rich, local or traditional community ecological knowledge with needed scientific knowledge.

... And in practice

Decentralized forest management examples cover a wide range of situations: forests devolved by governments to communities with community-based management systems or recognized as community-owned; community-managed hunting reserves; forested watersheds governed by local authorities with or without systems of payments for related environmental services; protected areas decentralized for local management through boards of stakeholders, communities or local government; customary agro-forestry systems or agro-pastoral systems adapted for decentralized forest management; sacred forests protected for religious and cultural value; forested territories of Indigenous Peoples; planted or natural forests assigned by local administrative authorities to individual household management; to name a variety. Noteworthy examples showcased in the literature are diverse in nature and scope – in their institutional arrangements and their relationship to central government authorities – including the community forests of Mexico and Central America, Nepal, Tanzania, Zambia and the Philippines, various Indigenous Peoples' forested territories in all geographic regions, individual household tenure in "collective responsibility" forests in China, and many traditional forests and agro-forests in Indonesia (see Box 8.1). In positive circumstances, community-managed forests have generated significant jobs and revenues for communities while increasing forest growth (both area and stocking) and providing multiple products and services of relevance to communities and the public (see examples in Molnar *et al.* 2007). In many instances, local communities are applying important innovations to adapt to the impacts of climate change and realign their management, conservation and production practices accordingly (Alcorn 2013; Sunderlin *et al.* 2014).

Developed countries are also decentralizing management. In the United States, private forest owners have been recognized as more efficient producers of timber and wood products, leading to a greater share of wood supply from private lands, greater stakeholder involvement in state and federal forest management decisions, and growing decentralization of community and municipal authority over forest management, conservation initiatives, and forest revenues. Additionally, Indigenous Peoples who have lost native tenure rights to the government are reclaiming these rights to access and manage or co-manage forests, and Spanish-descendant communities in the US Southwest are reasserting claims to national parks and federal and state forests, similar to social movements in the emerging economies.

There are a growing number of studies and meta-studies analyzing decentralized forest management and its advantages and disadvantages for achieving social and environmental goals. Research on the topic has increased in recent years due to the interest in REDD+ and climate change adaptation issues, and related questions about local resilience and capacity to respond to the impacts of climate change. Despite the growth in studies, definitive answers as to the effects of decentralization on forests and livelihoods are still lacking, for a number of reasons: many of the studies have been geographically restricted or heavily qualify their results (Seymour *et al.* 2014), or they have small sample sizes or have not implemented an experimental design including before/after or control/intervention site comparisons (Bowler *et al.* 2012); decentralized systems are evolving, so their effectiveness can change; policies that enable communities and other local authorities to effectively manage their natural resources may take years to implement or refine, thus judgment of effectiveness can often be premature (Hobley *et al.* 2012); and finally, studies use very different measures of success, making it hard to compare across studies and geographic regions (not to mention that Indigenous Peoples and forest communities often define success very differently from the government or outside agents).

Acknowledging these research limitations, several oft-cited studies have shown an empirical association between decentralized, local decision-making and tenure reform efforts and better forest condition. Chhatre and Agrawal (2009), in analyzing 80 tropical forest "commons", found that local autonomy and community ownership were both positively associated with increased carbon storage. Hayes and Persha (2010) and Persha *et al.* (2011) found that local rule-making autonomy or formal participation in rule-making led to better forest and biodiversity conservation outcomes. Under certain circumstances, locally managed forest areas resulted in better management outcomes than strictly protected areas (Bray *et al.* 2008; Nelson and Chomitz 2011; Soares-Filho *et al.* 2010), while also providing more local livelihood benefits, including for diverse categories of community members such as women and marginal groups (Agarwal 2010). Robinson *et al.* (2011) stress the importance of tenure *security*, over tenure type, with Pacheco (2009, 2012) and Larson *et al.* (2008) highlighting insecure forest tenure as one of the most significant barriers to sustainable forest use. Elson (2012) notes that tenure security appears to be correlated with people's willingness to make long-term investments and thus may help to explain better forest outcomes over the longer term. Importantly, Larson (2003) adds that securing tenure is insufficient to ensure good resource management. Outcomes also depend on other variables; particularly whether communities have economic incentives to conserve, or appropriate resources and capacity to assume responsibility.

While many studies have shown that decentralization of decision-making can benefit natural resource quality (Garnett *et al.* 2007; Sayer *et al.* 2008), in some cases these approaches have failed to deliver the expected and theoretical benefits to local communities as well as to the resource base (Blaikie 2006, in Malawi and Botswana). In cases in Cameroon, Brazil, India, the Philippines and Nepal, decentralization has perpetuated local tyrannies, fostered elite capture or marginalized particular groups (Edmunds and Wollenberg 2003; Oyono 2005; Timsina 2002; Pulhin 1996). These findings reinforce the idea that there is a role for central government to play, even within decentralized schemes, to check elite capture and other local inequalities. In an analysis of decentralized forest management in Bolivia, Andersson (2003) and Andersson *et al.* (2006) find that a key criterion for local institutions being effective managers is whether the central government provides regular backstopping and visits to the municipalities. A similar finding is documented in Nepal (Giri, 2012): where central government applied regulations encouraging equitable decision-making at the local forest management level, local institutions were more likely to address these issues over time. A nested set of governing institutions encourages the assumption of responsibility by municipal authorities and balances against skewed, local politics.

Which rights? Analyzing unsuccessful decentralization

A principal aspiration of decentralization is just and democratic governance that gives people a say in their own affairs (Agrawal and Gibson 1999). In some cases, effective decentralization has increased the capacity of local populations to make their needs and demands heard regionally and nationally, increasing the interactive capacity of local governments through fair elections, accountability mechanisms and local association (Larson 2003). Yet, this outcome is not assured. Despite the global trend towards devolution and/or local tenure recognition, particularly in lower- and middle-income countries, many cases have only resulted in a limited transfer of rights and authority, with continued over-regulation.

A more thorough analysis of the nature of the bundle of rights being transferred shows that a more limited set of rights are being recognized or transferred in the forest regimes governing these tenure arrangements, and that local tenure holders have trouble acting upon these new rights. In 27 forested countries in Africa, Asia and Latin America, for which comparable data exist regarding the bundle of rights, 31% of forest tenure regimes (19 of 61), the bulk of them in Latin America, include sufficient rights to be regarded as recognizing "ownership" of forest land by Indigenous Peoples and local communities. About 57% of the regimes (34 of 61) "designate" forests for communities, distributed fairly evenly among the three regions, and the remaining 12% (eight of 61), predominantly in Africa, remain "government administered" due to the weakness of the rights recognized. While four of 18 regimes created between 2002 and 2007 granted community ownership rights, none of the six created between 2008 and 2013 did so, instead only recognizing community control (five cases) or greater rights under government administration (one case) (RRI 2014). Central governments have often obstructed the decentralization process and retained control over resource management (Edmunds and Wollenberg 2003; Larson *et al.* 2008; Wittman and Geisler 2005). Ribot *et al.* (2006) note instances where governments have limited the kinds of powers transferred, thus undermining the ability of local governments to make decisions. The current situation is challenging. Thanh and Sikor (2006) note that actors can only hold actual power over management outcomes once legal rights, actual rights and practices mutually reinforce each other; this is still not a common situation.

Another source of failure is retaining historical but onerous rules and regulations, designed for government regulation of industrial-scale concessions with very different needs and incentives (Merino 2013). The majority of decentralized forest tenure regimes are heavily regulated with costly management plan requirements, high cost of documentation and permits, and the pervasive threat that government will transfer resource rights for commercial or development purposes to other jurisdictions or to the private sector. Counter-incentives can be created by other sectors, such as the incentives for agricultural settlers or customary rights-holders to clear forests as a means of demonstrating the value addition needed in agricultural laws to secure their tenure (de Jong *et al.* 2014; Capistrano 2008). Within the forest sector, there can be competing regulations for forest management and conservation, with one agency seeking to curtail use while another decentralizes authority to local people (Colfer and Capistrano 2005; Bray 2010). Forest departments are historically reluctant to relinquish control for fear of losing political power and their sector's share of the budget – heavier regulation provides a means for the forest department to justify their role and budget. Other challenges include the inability of governments to enforce their rules.

Which factors lead to positive outcomes?

Studies of the successes of decentralized forest management to some extent document a reverse image of the factors that lead to failure. Summarized here are some of the main lessons learned from decentralization that has led to positive social, economic and environmental outcomes.

Recognizing a sufficient part of the tenure bundle of rights to provide security and decision-making at community and local government levels is a principal success factor. Governments need to formalize local authority over decisions and tenure rights (Bray 2010), including the ability to exclude outsiders (RRI 2014). Regulatory frameworks need to be designed with the scale and nature of community forests in mind rather than imitate frameworks for industrial-scale commercial forestry (Hajjar *et al.* 2013; Gregerson and Contreras 2010; Fay and Michon 2005). Beyond a supportive business environment, a supportive enabling environment includes both technical assistance and capacity building of leaders and workers (Hernandez *et al.* 2010), and financial support, access to information, and higher political support. Too many reforms ignore the counter-incentives created by adverse policies outside the forest sector or fail to reform regulatory structures and the role of forest agencies.

It is also essential to provide decentralized institutions with time and space to develop new institutional capacity and evolve as they learn from experience (Hajjar *et al.* 2012). Forest management must be understood to be an adaptive process that will continue to change over time; failure to control deforestation and degradation at one point in time does not necessarily indicate that decentralized forest management will not be successful, but that effective models can only develop over time. Too often, communities are criticized for failing to achieve outcomes that state control has long failed to achieve. Oftentimes, nested institutions are needed where pressures are high – a balance of political and economic interests can only be achieved through multiple levels of governance. The state continues to play an important role in advancing decentralized forest management, but must be willing to devolve enough of a share of finances to enable local institutions to work.

Several cases have deviated from the "collective" management model. In Mexico, communities in some states such as Michoacán have long parceled out forest rights to individual households and harvesting decisions must therefore take into account this reality – modifying government harvesting prescriptions to allow for regular harvesting throughout the community forest (Merino 2013). In China, individual households make a more effective management unit; given the history of imposed collectivity in Han villages, the degree of desirable collective action in management varies by region and ethnic culture (Keliang and Riedinger 2011). In the Amazon, smallholder forest management has proved to be quite effective in managing for multiple social and environmental values (Padoch and Pinedo-Vasquez 2010; Pinedo-Vasquez *et al.* 2004). Each country is different and the lessons from one may not be appropriate for another, or even within the same country; contexts, histories and challenges of each country or region need to be considered in developing a strategy for supporting and implementing decentralized forest management. The diversity of successful decentralized management systems needs to be better understood, along with their cost/benefit streams.

Smart regulations are developed on the basis of the local reality, applying guiding principles but more cognizant of the constraints at the level of the community or local government, and of perverse pressures and incentives in the market and challenges to enforcement. One of the trends in the United States related to greater participation in public forest management is the switch in a number of states to best management practices. These are simple codes of conduct based on management objectives which are jointly established and monitored by stakeholders: government, loggers and mill owners, environmental organizations, and local forest owners and users. They require fewer government personnel to enforce, as monitoring is a joint responsibility with the other stakeholders, and reflect local interests and realities – and are much more likely to be enforced. Some developing countries have revised their frameworks. Nepal and Mexico eliminated stumpage fees on timber, realizing government would get greater revenues from value addition if they kept profits in the communities, and Mexico and Brazil have greatly simplified environmental and transport regulations (Gregerson and Contreras 2010; Bray 2010). However, in general, most countries need reforms to create an enabling environment for decentralized approaches to work for the benefit of both rural communities and forest conservation.

Way forward

A major challenge going forward is the limited progress in recent years in reforms. Fast-increasing pressures for forest acquisition or appropriation, particularly from urban development, commercial agriculture and sub-soil extraction of minerals, oil and gas, seriously threaten sound forest management and local institutions. These increasing pressures on the forest estate, exacerbated by poor accounting of environmental values and limited application of social and environmental safeguards, continue to lead many countries to draw down their natural forest capital, ignoring future resource values and the lost opportunity for the national economy and society.

Highlighted here are three elements of a way forward: the role of the state; the role of future products and services that can support local forestry; and a redefinition of what we mean by, and who defines, successful approaches to decentralization. Technical, business and market factors in support of local forest enterprises were discussed in Chapter 7.

The role of the state

Certainly the state has an important role going forward in enabling decentralized systems to evolve and in providing needed support. Primarily, governments need to respond to civil society demands to reform policies and regulations to level the playing field for Indigenous Peoples and communities in terms of access and use of their forests. Governments should continue to advance recognition of tenure rights where there are strong claims – community proposals exist in countries in the Amazon, sub-Saharan Africa (including the Congo) and Southeast and Insular Asia (including India and Nepal's lowlands and high mountains). Formalizing rights over land and resources needs to be sensitive to local power relations and realities, and any tenure reforms should be done with a better understanding of how gender, age, race and class dynamics shape local resource access and use. Often, a rush to formalize local rights has ignored such local dynamics, leading to further marginalization of vulnerable groups.

Governments also need to train forest agency staff and raise awareness and create incentives to change authoritarian attitudes. Can forest agencies (and the donors who support them) stop political foot-dragging in order to move reforms, support broad citizen participation and develop new skills related to environmental services and diversification of enterprise beyond timber to non-timber forest products (NTFPs) and bioenergy?

As stated above, nested approaches and a variety of governance arrangements can be important. For example, different co-management arrangements may make the best sense for high-value conservation areas, to share the burden of investment and protection with communities, and, in the case of environmental service markets and forest-based enterprises, provide assurances to private sector actors that value streams can be sustained over time.

Future forest products and services

Decentralized forest management can produce a wide variety of forest products and services, and decentralized forest managers will need to develop the right set of forest products and services, as well as the right market linkages. The diversity of choices include NTFPs, bioenergy and other less traditionally supported commodities, as well as growing ecosystem services markets. Decentralized managers should be encouraged to find their own balance. A mistake to be avoided is a dogged focus on promoting one type of forest enterprise – timber over NTFPs or vice versa – or tying tenure recognition to creation of a forest enterprise. Too many communities have started enterprises in Africa, Asia or Latin America because they saw no other way to gain legal recognition of their forest rights. Clearly their incentives to grow imposed enterprises will be limited at best.

A substantial number of community and smallholder enterprises worldwide already trade a variety of timber and non-timber products, but do so within the informal sector. While this has several advantages as well as disadvantages for communities and forests, many of these informal enterprises are being targeted by law enforcement efforts or villainized in national discourse, most recently by countries currently undergoing REDD+ or FLEGT reforms. It is important that reforms are adequately sophisticated in finding the best path forward for certifying "legal trade" without excluding or penalizing the poor and communities. Furthermore, efforts to link these small-scale enterprises to international markets may be misguided in many cases; as most of these enterprises already supply domestic markets, existing supply chains can be taken advantage of and enhanced rather than forcing ill-equipped small-scale enterprises to compete on international markets.

The wood-based bioenergy market – growing in importance globally and fast becoming a popular renewable energy source in forest-rich localities – can easily absorb small-scale and community plantations in addition to drawing on larger-scale plantations. The environment community and governments need to better assess this potential. The private sector can fruitfully explore bioenergy production at a community scale, in addition to investing in larger-scale models, including having corporate involvement in providing technical assistance. Of course, the private sector is highly diverse, and the established industry in many emerging economies has not often been a good partner to communities. As communities develop strong forest management, they will likely link to those existing or new private sector actors that are most sensitive to their interests and able to pursue shared goals. There is definitely a role for third parties (government or civil society based) to ensure such community–company partnerships are fair and equitable. Beyond bioenergy, many private sector actors are finding new models of business that lend themselves to the scale and nature of dealing with many small communities or municipalities. These include entering high-value markets for a changing range of value-added forest products produced at a smaller scale for a fast-growing middle class consuming culturally traditional products (medicinals, foods, fibers, furnishings) or newly popular natural or hand-crafted products. Branding that supports local forestry initiatives, such as community labeling, can give Indigenous Peoples and forest communities an edge in the marketplace, and forest certification bodies are currently exploring these options.

The importance of the role of communities and smallholders in any future REDD+ mechanism has been widely acknowledged, and more information is emerging on their role in climate resilience. Carbon payments will always be a relatively small part of total returns to decentralized forest management, so there need to be pay-offs for local people beyond carbon. Further, risk of local participation in carbon markets or schemes must be calculated and mitigated so that it is not a source of impoverishment, disempowerment or failure. Incorporation of local stakeholders, Indigenous Peoples, women and communities in rule-making bodies of international climate change mechanisms, such as the UN Framework Convention on Climate Change, the UN-REDD Programme and the Forest Carbon Partnership Facility, contributes to leveling the playing field, can help to shape these instruments away from the narrow interest of sovereign states, and educates the international donor community on local needs and concerns. The international community needs to continue to reform conservation models to be truly rights-based, making local communities and Indigenous Peoples full participants in visioning, design, implementation, monitoring and benefit sharing.

Water flow and quality is an additional environmental service that will become ever more important in the future; expanded new service markets can support decentralized forest managers as well as create upstream and downstream incentives for management.

Redefining goals

Measures of success of decentralization efforts need to look at a complex set of goals and values, focusing on local as well as national or global ones. In too many community-based conservation and management projects, conservation or carbon storage has trumped local social, cultural, political and economic considerations. This is simply not sustainable; contributions to poverty reduction (using a holistic definition of "poverty" and not just monetary income), local employment, and political and cultural empowerment need to be paramount for such efforts to bear fruit over time.

Furthermore, it is essential that all actors better understand how traditional livelihood systems – shifting cultivation, agro-forestry and agro-pastoral systems – can be integrated with forest management, and what the implications of this integration are for ecological and social benefits to and impacts on forests and communities. Better understanding the spatial, temporal and economic integration of these livelihood practices can help to modify the western models of introduced community forestry practices, and gain relevance, effectiveness and consistency with internationally agreed environmental and human rights. Nobel prize winner Elinor Ostrom's "Ostrom's Law" can be used as a guide: to paraphrase, if it works in practice, it can work in theory (Fennell 2011). There are innumerable examples around the world of self-generated community and local natural resource management; introduced initiatives need to refocus on these "discovered" self-generated examples, find out what works in practice, and design technical assistance and capacity building appropriately, and tailored to local realities and interests.

Decentralization is a process of social and political transformation with several advantages to centralized control, assuming enabling conditions are put in place and multi-level, nested institutions adequately check and balance local authority. Advantages include: incentives for reduced carbon emissions from deforestation and degradation; generating other key environmental services; supporting livelihood and income streams with greater social inclusion over time; and local adaptation and innovation to address impacts of climate change. Maximizing benefits to the local economy by integrating forest resource management with agricultural, pasture and agro-forestry systems can create stronger, more lasting incentives for good management.

Yet, decentralization of forest management faces big challenges. Progress in community tenure recognition has slowed down considerably in the last five years; the area secured between 2008 and 2013 is less than 20% of the area secured in the previous six years. Furthermore, what has been recognized is less and less of the complete rights bundle – unless governments provide security to decentralized authorities, management and enterprise benefits will be reduced. However, we believe that decentralization of natural resource and forest management will continue in Africa, Asia and Latin America, despite the uneven progress of reforms, given demands from communities and civil society for local tenure and rights recognition, continuing pressure from globalization and modernization for governance reforms in emerging economies, and the growing evidence that locally controlled forestry can be better for checking deforestation and conserving environmental values more equitably, particularly in the face of climate change.

Box 8.1

Globally, there are many examples of decentralized forest management through communities and other forms of local governance. These examples are quite diverse in their history of emergence, nature, scope, institutional arrangements, relationships to government and maturity. Here we briefly present a few examples of models being pursued in different countries; by no means is this an exhaustive list of models globally, or even within the countries mentioned.

Mexico is seen as having one of the most advanced community forestry sectors in the world. Some 60–80% of forests in Mexico are owned by communities, and several hundred community-based forestry enterprises in operation have helped to curb deforestation and combat rural poverty. Since the 1980s, communities have been given progressively more rights over their forests, and a number of successive government policies have enabled the development of a community forestry sector. While initially focused on timber, in the more successful examples communities have diversified their enterprises to include non-timber forest products and services, including commercial spring water bottling, pine resin and ecotourism. However, these examples do not represent the majority of communities, where many remain under-skilled and under-capitalized, or still struggle to combat illegal logging. Excessive regulation and taxation have also been blamed for holding back the community forest sector, particularly as competition from plantation wood from South America continues to increase.

Decentralized forest management in **Brazil** over the years came about from a combination of indigenous and non-indigenous social movements pushing for local rights over resources, and an exogenous push mostly from NGOs trying to promote more sustainable forest practices in the Amazon. Community-based management models vary greatly within the country, with some under the jurisdiction of extractive reserves, territories, forestry settlement areas and sustainable development areas, to name a few. Other local management arrangements, particularly smallholder-based ones, continue to occur informally. The problems faced by many Brazilian forest-dependent communities are familiar to other regions of the world: insecure or unregularized tenure in many parts, lack of access to credit, markets and technical assistance, and weak social organization.

In 1998, a half million hectares of forests in the buffer zone of the tropical Mayan biosphere reserve, in **Guatemala**, were transferred in a 20-year contract to 23 communities (two Indigenous Peoples and 21 mestizo colonists), in response to an active social movement of colonists who refused to be driven out from the area by the creation of the reserve. The communities and their support organization, ACOFOP, have become relatively independent of external NGO oversight, and have heavily protected the reserve while developing successful timber and non-timber based enterprises. At first, they sold high-value sawn timber and some finished furniture and wood mats, domestically and to Europe, and have since diversified into ornamental palms, forest foods and condiments. Women have been increasingly involved in enterprise management and small entrepreneurship. Community forests continue to face pressures from the agriculture frontier and from archaeologists and politicians who want to expropriate large parts of the reserve to develop high-end tourism. Communities still have no assurance of extension of their contract or granting of greater rights, despite their annual investment of more than US$160,000 in fire control and daily protection of forest boundaries. Like elsewhere, decentralization of forest management has led to social and political transformation in the communities.

In **The Gambia**, experiments with decentralization beginning in the 1990s were a response to the failure of the state to manage the country's forests effectively. Community forests were established as a part of this, where communities would enter into an agreement with the government to manage forests within their traditional land. A phased approach is taken where the government plays a key role at first in providing business and technical training to communities, as well as economic support in the form of micro-loans and start-up materials. In the final phase, the community acquires exclusive rights over the forest resources, after having demonstrated their capacity to manage the forest sustainably. Both community-controlled and individually owned businesses can operate within the confines of a community forest, with the community-owned businesses resulting in much needed social investments. Despite successes with this approach to date, the substantial upfront investments by the government has resulted in only limited expansion, covering just 6% of Gambia's forests.

Tanzania began its decentralization of forest management using the Joint Forest Management (JFM) model, where the government established lease agreements with local communities for forest areas, but retained control of key decision-making around management and harvesting, and retained the right to withdraw the lease. This was problematic in part because only the less-productive forests were handed to villagers, limiting the economic incentives for management, and also because the guidelines for establishing JFM agreements were never approved due to lack of consensus about the cost and benefits sharing mechanism. As a consequence, it is very difficult to implement a JFM in practice. Starting in the 1990s, Tanzania created a participatory forest management regime under which the bundle of rights has been more robust, with more control over access and use but still limited exclusion rights. Studies have shown that these regimes have been better managed than JFM sites.

China is an example of extensive decentralization of forest management. Forest land reforms initially focused on devolving forests from central to collective management at village level, but this had mixed success. In 2003, a new policy, complemented by other rural reforms, empowered collectives to delegate formal forest control to individual households. Families continue to group together for certain services or access to market, but plantation parastatals have largely disappeared and members of collectives sell timber directly to mills and other forest industries. Over 102 million hectares have been transferred, involving 400 million people, with significant increases in household income in the districts and provinces concerned and increased afforestation rates in villages with forests devolved to households in comparison with villages where forest management remains collective.

Nepal revised its forest law in 1978 to address dire environmental and economic problems related to deforestation and degradation by decentralizing management authority to local people. The law made it possible for communities to form forest user groups for specific areas of forest traditionally within their use domain and/or customarily owned. Since then, 1.6 million households have been involved in over 17,500 community forest user groups, managing nearly 23% of the nation's forests. This has led to increased forest cover that has improved water flow and quality and reduced landslides and disaster risks. Despite limited enterprise development, the forest user groups have led to a modest increase in employment and household incomes, much greater security of access to non-timber and timber products used in daily livelihoods, increased participation in village development plans and decisions, and less discrimination against marginal groups and women, though inequality continues to be an issue.

There are many examples of decentralized forest management in **Indonesia**, ranging from traditional forests of Indigenous Peoples to smallholder and agroforestry on the densely populated island of Java. Smallholder plantations and agro-forestry on the island of Java supply the bulk of teak and mahogany to domestic furniture-making and other factories, which in turn sell their products on international markets. In Central Java Province, wood furniture is the largest contributor to provincial exports, accounting for 27.16% of total exports in 2000. Some of these smallholders on Java and outer islands have grouped together to be certified by the Forest Stewardship Council, after having received training and support from international NGOs, and have the potential to be highly competitive if their tenure and rights are adequately recognized.

Responding to years of public pressure, the provincial government of British Columbia in **Canada** began a pilot program of community forestry tenures in the late 1990s to increase the participation of communities and First Nations in the management of local forests for local benefits. By 2004, the government established the Community Forest Agreement as a permanent program. Community forest governance structures have taken a wide variety of forms, from cooperatives to community-owned corporations and First Nations bands. However, the province has been criticized for creating an institutional arrangement that does not differ markedly from the industrial model, and with limited devolution of authority. Despite a steady increase in the number of communities involved in the program over the last decade, community forest tenures still only account for a very small percentage of British Columbia's forests.

Note

1 Statutory entails that it is recognized by the formal legal system of a country.

References

Agarwal, B., 2010. *Gender and Green Governance: The Political Economy of Women's Presence within and beyond Community Forestry*. Oxford, UK: Oxford University Press.

Agrawal, A. and Gibson, C., 1999. Enchantment and disenchantment: the role of community in natural resource conservation. *World Development*, 27(4), pp. 629–649.

Agrawal, A. and Ribot, J.C., 1999. Accountability in decentralization: a framework with South Asian and West African cases. *The Journal of Developing Areas*, 33(4), pp. 473–502.

Agrawal, A., Chhatre, A. and Hardin, R., 2008. Changing governance of the world's forests. *Science*, 320, pp. 1460–1462.

Alcorn, Janis B., 2013. *Lessons Learned from Community Forestry in Latin America and Their Relevance for REDD+*. Washington, DC: USAID-supported Forest Carbon, Markets and Communities (FCMC) Program.

Andersson, K., 2003. What motivates municipal governments? Uncovering the institutional incentives for municipal governance of forest resources in Bolivia. *Journal of Environment and Development*, 12(1), pp. 5–27.

Andersson, K.P., Gibson, C.C. and Lehoucq, F., 2006. Municipal politics and forest governance: comparative analysis of decentralization in Bolivia and Guatemala. *World Development*, 34(3), pp. 576–595. doi:10.1016/j.worlddev.2005.08.009.

Barsimantov, J.*et al.*, 2011. When collective action and tenure allocations collide: outcomes from community forests in Quintana Roo, Mexico and Petén, Guatemala. *Land Use Policy*, 28(1), pp. 343–352.

Berkes, F., Colding, J. and Folke, C., 2003. *Navigating Social–Ecological Systems: Building Resilience for Complexity and Change*. Cambridge: Cambridge University Press.

Blaikie, P., 2006. Is small really beautiful? Community-based natural resource management in Malawi and Botswana. *World Development*, 34(11), pp. 1942–1957.

Bowler, D.E.*et al.*, 2012. Does community forest management provide global environmental benefits and improve local welfare? *Frontiers in Ecology and the Environment*, 10(1), pp. 29–36.

Bradshaw, B., 2003. Questioning the credibility and capacity of community-based resource management. *The Canadian Geographer*, 47(2), pp. 137–150.

Bray, D.*et al.*, 2008. Tropical deforestation, community forests, and protected areas in the Maya forest. *Ecology and Society*, 13(2), p. 56.

Bray, D.B., 2010. *Putting the community back into community forestry: The enchantment of collective action for timber production in Latin America*. CIFOR "Taking stock of smallholder and community forestry" Montepellier, France. www.slideshare.net/CIFOR/putting-the-community-back-into-community-forestry-the-enchantment-of-collective-action-for-timber-production-in-latin-america.

Brown, H.*et al.*, 2010. Institutional choice and local legitimacy in community-based forest management: lessons from Cameroon. *Environmental Conservation*, 37(3), pp. 261–269.

Capistrano, D., 2008. Decentralization and forest governance in Asia and the Pacific: trends, lessons and continuing challenges. In C. Colfer, G. Ram Dahal and D. Capistrano, eds. *Lessons from Forest Decentralization: Money, justice and the quest for good governance in Asia-Pacific*. London: Earthscan.

Chapela, F., 2005. Indigenous community forest management in the Sierra Juarez, Oaxaca. In D. Bray, L. Merino-Perez and D. Barry, eds. *The Community Forests of Mexico: Managing for sustainable landscapes*. Austin: University of Texas Press.

Charnley, S. and Poe, M.R., 2007. Community forestry in theory and practice: where are we now? *Annual Review of Anthropology*, 36(1), pp. 301–336.

Chhatre, A. and Agrawal, A., 2009. Trade-offs and synergies between carbon storage and livelihood benefits from forest commons. *Proceedings of the National Academy of Sciences*, 106(42), pp. 17667–17670.

Colchester, M., 2004. Conservation policy and indigenous peoples. *Environmental Science and Policy*, 7(3), pp. 145–153.

Colchester, M., 2008. *Beyond Tenure: Rights-based approaches to peoples and forests – Some lessons from the Forest Peoples Programme*. Washington, DC: Rights and Resources Initiative, pp.1–52.

Colfer, C. and Capistrano, D. (ed.), 2005. *The Politics of Decentralization: Forests, people and power*. Center for International Forestry Research, UK and US: Earthscan Library.

Contreras-Hermosilla, A.*et al.*, 2006. *Forest Governance in Countries with Federal Systems of Government: Lessons for decentralization*. Bogor, Indonesia.

Crona, B.*et al.*, 2011. Combining social network approaches with social theories to improve understanding of natural resource governance. In O. Bodin and C. Prell, eds. *Social Networks and Natural Resource Management: Uncovering the fabric of environmental governance*. Cambridge: Cambridge University Press.

de Jong, W. *et al.*, 2014. Opportunities and Challenges for Community Forestry. In M. Gerardo, P. Katila, G. Galloway, R. Alfaro, M. Kanninen, M. Lobovikov and J. Varjoin, eds. *IUFRO World Series Vol. 25: Forests and Society – Responding to global drivers of change*, pp. 299–314.

Edmunds, D. and Wollenberg, E., 2003. *Local Forest Management: The impacts of devolution policies*. London: Earthscan.

Elson, D., 2012. *Guide to Investing in Locally Controlled Forestry*. London: Growing Forests Partnership.

Fay, C. and Michon, G., 2005. Redressing forestry hegemony: when a forestry regulatory framework is best replaced by an agrarian one. *Forests, Trees and Livelihoods*, 15, pp. 193–209.

Fennell, L., 2011. Ostrom's law: property rights in the commons. *International Journal of the Commons*, 5(1), pp. 9–27.

Garnett, S.T., Sayer, J. and Du Toit, J., 2007. Improving the effectiveness of interventions to balance conservation and development: a conceptual framework. *Ecology and Society*, 12(1), p. 2.

Giri, K., 2012. *Gender in forest tenure: Pre-requisite for sustainable forest management in Nepal*. The Challenges of Securing Women's Tenure and Leadership for Forest Management: The Asian Experience. Brief#1. Washington DC: Rights and Resources Initiative.

Gregersen, H. and Contreras, A., 2010. *Rethinking Forest Regulations: from simple rules to systems to promote best practices and compliance*. Washington DC: Rights and Resources Initiative.

Growing Forests Partnership, n.d. *The Forests Dialogue: Investing in Locally Controlled Forestry* (ILCF).

Hajjar, R.F., Kozak, R.A. and Innes, J.L., 2012. Is decentralization leading to "real" decision-making power for forest-dependent communities? Case Studies from Mexico and Brazil. *Ecology and Society*, 17(1), p. 12.

Hajjar, R.*et al.*, 2013. Community forests for forest communities: integrating community-defined goals and practices in the design of forestry initiatives. *Land Use Policy*, 34, pp. 158–167.

Hayes, T. and Persha, L., 2010. Nesting local forestry initiatives: revisiting community forest management in a REDD+ world. *Forest Policy and Economics*, 12(8), pp. 545–553.

Hernandez, T., Fortin, R. and Butterfield, R., 2010. *Impacts of Technical Assistance on a Community Forest Enterprise: the case of San Bernardino de Milpillas Chico, Mexico*, Rainforest Alliance, p. 6.

Hobley, M.*et al.*, 2012. *Persistence and Change: Review of 30 years of community forestry in Nepal*. Kathmandu: Multi Stakeholder Forestry Programme (MSFP), www.msfp.org.np.

Keck, M.E., 1995. Social equity and environmental politics in Brazil: lessons from the rubber tappers of Acre. *Comparative Politics*, 27(4), pp. 409–424.

Keliang, Z. and Riedinger, J., 2011. "Chinese Farmers' Land Rights at the Crossroads – Findings and Implications from a 2010 Nationwide Survey". Presentation for World Bank Conference on Land and Poverty, 18 April 2011, Washington, DC.

Kellert, R.*et al.*, 2000. Community natural resource management: promise, rhetoric, and reality. *Society and Natural Resources*, 13(8), pp. 705–715.

Larson, A.M., 2003. Decentralisation and forest management in Latin America: towards a working model. *Public Administration and Development*, 23, pp. 211–226.

Larson, A., 2005. Democratic decentralization in the forestry sector: lessons learned from Africa, Asia and Latin America. In C. Colfer and D. Capistrano, eds. *The Politics of Decentralization: Forests, power and people*. London: Earthscan.

Larson, A. and Dahal, G., 2012. Forest tenure reform: new resource rights for forest-based communities? *Conservation and Society*, 10(2), p. 77.

Larson, A.M.*et al.*, 2008. *Tenure rights and beyond: Community access to forest resources in Latin America*. CIFOR Occasional Paper, p. 104.

Manor, J., 1999. *The Political Economy of Democratic Decentralization*. Washington, DC: The World Bank, pp. 1–145.

McCarthy, J., 2006. Neoliberalism and the politics of alternatives: community forestry in British Columbia and the United States. *Annals of the Association of American Geographers*, 96, pp. 84–104.

Merino, L., 2013. Conservation and forest communities in Mexico. An ongoing struggle over forest property rights? In L. Porter-Bolland, I. Ruiz-Mallen, C. Camacho-Benavides and S. McCandless, eds. *Community Action for Conservation: Mexican experience*. New York: Springer. Spanish original: *Encuentros y Desencuentros. La Política Forestal en Tiempos de Transición Política en México* (co-authored with G . Ortiz). Mexico: Instituto de Investigaciones Sociales, UNAM and Miguel Ángel Porrúa Editorial.

Molnar, A., Scherr, S. and Khare, A., 2004. *Who Conserves the World's Forests? A new assessment of conservation and investment trends*, Washington, DC: Forest Trends.

Molnar, A.*et al.*, 2007. *Community-based Forest Enterprises in Tropical Forest Countries: Status and potential*. ITTO, Rights and Resources Initiative and Forest Trends, p. 205.

Molnar, A.*et al.*, 2010. *Small Scale, Large Impacts: Transforming Central and West African sustainable development, growth, and governance*, Washington, DC: Rights and Resources Initiative.

Nelson, A. and Chomitz, K.M., 2011. Effectiveness of strict vs. multiple use protected areas in reducing tropical forest fires: a global analysis using matching methods. *PloS one*, 6(8), p.e22722.

Oyono, P.R., 2004. One step forward, two steps back? Paradoxes of natural resources management decentralisation in Cameroon. *The Journal of Modern African Studies*, 42(1), pp. 91–111.

Oyono, P.R., 2005. Profiling local-level outcomes of environmental decentralizations: the case of Cameroon's forests in the Congo Basin. *The Journal of Environment and Development*, 14(3), pp. 317–337.

Pacheco, P., 2009. Agrarian reform in the Brazilian Amazon: its implications for land distribution and deforestation. *World Development*, 37(8), pp. 1337–1347.

Pacheco, P., 2012. Smallholders and communities in timber markets: conditions shaping diverse forms of engagement in tropical Latin America. *Conservation and Society*, 10(2), p. 114. doi:10.4103/0972-4923.97484.

Padoch, C. and Pinedo-Vasquez, M., 2010. Saving slash-and-burn to save biodiversity. *Biotropica*, 42(5), pp. 550–552.

Pagdee, A., Kim, Y. and Daugherty, P., 2006. What makes community forest management successful: a meta-study from community forests throughout the world. *Society and Natural Resources*, 19(1), pp. 33–52.

Pahl-Wostl, C., 2009. A conceptual framework for analysing adaptive capacity and multi-level learning processes in resource governance regimes. *Global Environmental Change*, 19(3), pp. 354–365.

Persha, L., Agrawal, A. and Chhatre, A., 2011. Social and ecological synergy: local rulemaking, forest livelihoods, and biodiversity conservation. *Science*, 331(6024), pp. 1606–1608.

Pinedo-Vasquez, M.*et al.*, 2004. Post-boom logging in Amazonia. *Human Ecology*, 29(2), pp. 219–239.

Pulhin, J., 1996. *Community forestry: paradoxes and perspectives in development practice*. PhD Dissertation, Australian National University.

Raik, D., Wilson, A. and Decker, D., 2008. Power in natural resources management: an application of theory. *Society and Natural Resources*, 21(8), pp. 729–739.

Ribot, J., 2004. *Waiting for democracy: The politics of choice in natural resource decentralization*. World Resource Institute Report, pp. 1–154.

Ribot, J.C., Agrawal, A. and Larson, A.M., 2006. Recentralizing while decentralizing: how national governments reappropriate forest resources. *World Development*, 34(11), pp. 1864–1886.

Robinson, B., Holland, M. and Naughton-Treves, L., 2011. *Does Secure Land Tenure Save Forests? A review of the relationship between land tenure and tropical deforestation*, Copenhagen: CGIAR Research Program on Climate Change, Agriculture and Food Security (CCAFS).

RRI, 2014. *What Future for Reform? Progress and slowdown in forest tenure reform since 2002*. Washington, DC: Rights and Resources Initiative.

Samoff, J., 1990. Decentralization: the politics of interventionism. *Development and Change*, 21(3), pp. 513–530.

Sayer, J.*et al.*, 2008. *Local Rights and Tenure for Forests: Opportunity or threat for conservation?* Washington, DC: Rights and Resources Initiative.

Scherr, S.J., White, A. and Kaimowitz., D., 2004. *Making Markets Work for Forest Communities*. Washington, DC: Forest Trends, pp. 130–155.

Schlager, E. and Ostrom, E., 1992. Property-rights regimes and natural resources : a conceptual analysis. *Land Economics*, 68(3), pp. 249–262.

Schroeder, R., 1999. Community, forestry and conditionality in the Gambia. *Africa*, 69(1), pp. 1–22.

Seymour, F., La Vina, T. and Hite, K., 2014. *Evidence Linking Community-Level Tenure and Forest Condition : An Annotated Bibliography*. San Francisco, CA: Climate and Land Use Alliance.

Sharma, C.K., 2006. Decentralization dilemma: measuring the degree and evaluating the outcomes. *Indian Journal of Political Science*, 67, 49–64.

Soares-Filho, B.*et al.*, 2010. Role of Brazilian Amazon protected areas in climate change mitigation. *Proceedings of the National Academy of Sciences*, 107(24), pp. 10821–10826.

Sunderlin, W. *et al.*,2014. How are REDD+ proponents addressing tenure problems? Evidence from Brazil, Cameroon, Tanzania, Indonesia, and Vietnam. *World Development*, 55: 37–52.

Tacconi, L., 2007. Decentralization, forests and livelihoods: theory and narrative. *Global Environmental Change*, 17(3–4), pp. 338–348.

Thanh, T. and Sikor, T., 2006. From legal acts to actual powers: devolution and property rights in the Central Highlands of Vietnam. *Forest Policy and Economics*, 8, pp. 397–408.

Timsina, N., 2002. Empowerment or marginalization: a debate in community forestry in Nepal. *Journal of Forest and Livelihood*, 2(1), pp. 27–33.

Wittman, H. and Geisler, C., 2005. Negotiating locality: decentralization and communal forest management in the Guatemalan highlands. *Human Organization*, 64(1), pp. 62–74.

9 Promises and perils of plantation forestry

Jacki Schirmer, Romain Pirard and Peter Kanowski

Plantation forestry is an increasingly important part of the global forestry industry. In the decade to 2010, global expansion of planted forests averaged five million hectares (ha) annually, mostly through afforestation of land that had not been recently forested (FAO 2010). This rate of expansion is substantially higher than in previous decades (FAO 2010) and is happening worldwide, as shown in Table 9.1, although more than half of all planted forests are located in just five countries (China, the United States, the Russian Federation, Japan and India). Within this global snapshot, there are very different trends: countries such as China (Xu 2011) and Vietnam (Government of Vietnam 2011) are pursuing rapid growth of tree plantations, while in much of North America and Europe there is a large existing plantation estate and more limited expansion (FAO 2010).

Of the estimated 264 million ha of planted forests established globally by 2010, around three-quarters had commercial wood production as their predominant purpose (FAO 2010). Plantations are of growing importance to global timber supply, in part because of declining wood production from natural forests (Warman 2014). Plantations are estimated to contribute, currently and for the foreseeable future, around one-third of industrial wood supply (Barua *et al.* 2014; Jurgensen *et al.* 2014). The supply of industrial roundwood from plantations could increase to 1.5 billion m^3 in 2050, but even at this volume is expected to contribute only around one-third of global wood supply (Barua *et al.* 2014).

Depending on the scenarios used, planted forests are predicted to cover anywhere from 303 to 345 million ha by 2030, with most of the absolute increase taking place in Asia, as can be seen in Table 9.1 (Carle and Holmgren 2008; Warman 2014). Plantations established for industrial roundwood production are likely to dominate, although plantations are increasingly being established for other commercial purposes, with an estimated 403,000 ha of plantations established for commercial carbon sequestration worldwide by September 2011 (Diaz *et al.* 2011), and a growing area of plantations established for biofuel production (Pin Koh and Ghazoul 2008; Arevalo *et al.* 2014).

Table 9.1 Where are the plantations? Area and rate of growth of planted forests by region and country

Region/country	*Area of planted forest, 2010 (1 000 ha)*	*Average annual rate of afforestation, 2005*[a] *(1000 ha/yr)*	*Average annual change in area of planted forest 2005–2010*[a] *(1000 ha/yr)*
South Africa	1763		3
Eastern and Southern Africa	**4116**	**49**	
Sudan	6068		43
Northern Africa	**8091**	**53**	
Western and Central Africa	**3203**	**48**	
Total Africa	**15409**	**150**	
China	77157		1988
Japan	10326		No data
Republic of Korea	1823		0.5
East Asia	**90232**	**4385**	
India	10211		145
Indonesia	3549		−30
Malaysia	1807		47
Thailand	3986		108
Vietnam	3512		144
South and Southeast Asia	**25552**	**398**	
Turkey	3418		160
Western and Central Asia	**6991**	**142**	
Total Asia	**122775**	**4,926**	
Belarus	1857		20
Czech Republic	2635		2
Finland	5904		0
France	1633		5
Germany	5283		0
Hungary	1612		9
Norway	1475		15

Region/country	*Area of planted forest, 2010 (1 000 ha)*	*Average annual rate of afforestation, 2005*[a] *(1000 ha/yr)*	*Average annual change in area of planted forest 2005–2010*[a] *(1000 ha/yr)*
Poland	8889		24
Romania	1446		8
Russian Federation	16991		6
Spain	2680		26
Sweden	3613		0
Ukraine	4846		12
United Kingdom	2219		6
Total Europe	**69318**	**169**	
Caribbean	**548**	**35**	
Central America	**584**	**4**	
Canada	8963		183
Mexico	3203		162
United States of America	25363		188
North America	**37529**	**199**	
Total North and Central America	**38661**	**204**	
Australia	1903		55
New Zealand	1812		-8
Total Oceania	**4101**	**59**	
Argentina	1394		38
Brazil	7418		331
Chile	2384		64
Total South America	**13821**	**88**	
World	**264084**	**5,595**	

Data in this table are for all types of planted forests, including those planted for non-commercial purposes. FAO (2010) estimated that approximately three-quarters of these had commercial timber production as their primary purposes. Only countries with greater than 1 million ha of planted forests in 2010 are included

[a] The FAO have a slightly different definition and calculation for (i) afforestation rates versus (ii) average annual change in area. These statistics are therefore presented in separate columns (the FAO have not produced each type of data for both global regions and individual countries)

Data source: FAO (2010)

There is ongoing debate about the sustainability of tree plantations, making the plantation sector a critical area to consider when examining the opportunities and threats presented by the business activities of the global timber industry. Ensuring sustainability is not only critical to ensuring positive environmental, economic and social outcomes: it is increasingly a critical component of even short-term business viability, particularly as businesses face greater pressure from consumers and risk losing their 'social licence to operate' if they are perceived to be operating unsustainably.

In this chapter, we consider the promises and perils of plantations grown for commercial timber production, where 'timber' is defined as all forms of roundwood production for commercial purposes. We chose this focus as the global plantation estate is dominated by commercial timber plantations, and there is a larger body of evidence regarding sustainability for these plantations than for other commercial tree plantations, such as those grown for carbon sequestration and biofuel production. We do not examine planted forests established for non-commercial purposes such as addressing environmental degradation, in order to better focus on the tensions between commercial business practice and sustainability.

Business practices in the plantation sector influence sustainability at many points and in many ways. We do not attempt to review all sustainability concerns related to plantations – a task impossible in a single chapter – but focus on key areas that must be considered if timber production from tree plantations is to be a truly sustainable business enterprise. First, we review the debate that has emerged around sustainability of plantations, identifying common sustainability concerns that regularly emerge. We then examine three prominent sustainability challenges in more detail, each of which has emerged in multiple countries: land ownership and tenure rights; deforestation and biodiversity; and employment and communities. Following this we consider the design of plantations and associated governance systems to enable sustainability. First, we examine arguments surrounding the scale of plantations, focusing on whether a shift from large-scale to small-scale plantations is likely to address sustainability concerns. We then consider enabling conditions – the broader social and political conditions that enable the success or failure of plantations from both a business and a sustainability perspective. We conclude by considering future directions: how can appropriate governance and business approaches ensure a successful plantation sector from the point of view of both ensuring viable businesses, and of enabling sustainable development?

Plantations and sustainability: A contentious debate

Plantation forestry is widely promoted for the business opportunities it provides to the timber industry through high rates of wood production and the ability to produce consistent quality timber (Rudel 2009). Plantations are also argued to have the potential to address problems as diverse as rural poverty, environmental degradation, deforestation and climate change. For example, it is argued that plantations can help address: rural poverty through providing employment and income earning opportunities in rural communities; environmental degradation through reducing logging pressure on natural forests; and climate change through sequestering carbon (Righelato and Spracklen 2007).

However, this positive view of plantation forestry is not shared by all: critics argue that many plantations cause environmental, economic and social damage to rural landscapes and communities. Contention and conflict over plantation expansion has been documented in more than 35 countries, including all of those in which large areas of plantation have been established (Schirmer 2007; Mola-Yudego and Gritten 2010; Gerber 2011). Concerns have been raised about whether local communities receive the economic benefits of plantations; about the environmental implications of planting large areas of a single species; and about whether plantations truly provide the economic returns often promised, to cite just a few examples.

Even the terminology used to discuss tree plantations is contentious, as it helps define what constitutes a plantation. Some argue that the word 'forest' should not be associated with tree plantations, as this may imply they have all the characteristics of natural forests and hence can replace them. At the other end of the spectrum, some use terms such as 'tree engineering' to emphasize the technological and man-made component of plantations. To resolve these issues – and for statistical purposes as well – a new term was proposed by the FAO: planted forests (Del Lungo *et al.* 2006). It covers a range of ecosystems from semi-natural forests where trees were planted, to strictly man-made tree plantations.

In this chapter, we use the terms 'tree plantation', 'plantation forestry' and 'planted forest' interchangeably, together with more specific terms: 'timber plantations' refers to any plantation that produces roundwood for commercial wood and paper production; 'industrial plantations' refers to those established in large areas, managed by a company and aimed at supplying large volumes of roundwood to wood or paper processing facilities; while 'smallholder plantations' refers to small-scale plantations, owned by individual landholders or small cooperatives, which may supply a range of markets from small-scale local sale to large-scale processors.

Debates about the benefits and costs of tree plantations often focus on the sustainability of those plantations – in other words, whether the rapidly growing global plantation estate is being established and managed in a way that 'meets the needs of the present without compromising the ability of future generations to meet their own needs' (World Commission on Environment and Development 1987). This requires plantations to be sustainable in terms of their environmental, economic and social impacts, which in turn requires businesses to go beyond focusing on the shorter-term needs of their business to considering issues such as the biodiversity impacts of plantations, the economic flow-on effects of plantation forestry, equity in the distribution of benefits from the use of land, and access to employment and income earning opportunities. Without this focus on sustainability, business practices can give priority to short-term benefits and high timber production, a focus that can come at the expense of long-term productive use of land and generate negative externalities, such as reduced access to forest products, loss of land for local communities or pollution.

Common debates that have emerged in recent decades about sustainability of plantations are summarised in Table 9.2. Perhaps the most evident fact in Table 9.2 is that polarised views are common, with views about the effects of plantations often contradicting each other. We argue that making global claims about the sustainability of plantations is unwise and unhelpful. In reality, the sustainability or otherwise of a given tree plantation depends on many factors, including the economic relationships involved, the type and location of plantation, previous land uses, the quality of the management of the plantation, quality of governance systems that are in place to oversee planted forests, and of course the unique characteristics of the local communities and landscapes in which the plantation is established. This approach recognises that plantations are not inherently sustainable or unsustainable: their effects depend on the actions of those involved in establishing, managing and governing them.

The importance of examining the specific context and understanding how sustainability of plantations varies even within the same region can be seen by examining the case studies of plantation expansion described in the following pages. In the first, expansion of plantations in the United Kingdom in the 1980s was criticised with regard to environmental sustainability when a combination of factors led to plantations being established on drained areas of peatland. In the second, large-scale plantations in Indonesia have been criticised on environmental, social and economic fronts while small-scale plantations have not been criticised on the sustainability front, but have failed in terms of business success. In the third, however, small-scale plantations in Thailand have achieved reasonable success in terms of providing both a livelihood for thousands of smallholders and also providing a supply of timber supporting a large-scale pulp industry, while receiving little criticism regarding sustainability-related issues, despite an earlier history of conflict over large-scale plantations in the country. These three case studies were chosen because they demonstrate common sustainability opportunities and challenges that have emerged in different countries, together with varying levels of 'business' success. They were also chosen because they are well documented – something that remains a rarity, despite the widespread literature variously criticising and promoting plantations. For example, despite widespread discussion of the importance of understanding sustainability of plantations in China, a country where some of the largest scale and most rapid plantation expansion is occurring, there remains little robust, documented research that documents and critically examines claims about many aspects of sustainability of China's large areas of tree plantations. This and

Table 9.2 Commonly cited benefits and costs of tree plantations

Topic	*Benefits*	*Costs*
Land tenure and land markets	Plantations provide a different market for land, which may boost land prices and benefit landholders seeking to sell or lease out their land	Previous land users may be displaced by plantations without compensation or recognition of their claim to their land; or unfair deals for land may disadvantage those who are selling or leasing the land. Plantation expansion may result in increasing land prices that price other land users, such as farmers, out of the market
Employment opportunities	Proponents argue that plantations generate employment opportunities for local people in establishing, managing, harvesting and processing plantations. This is argued to help stem decline in rural populations and provide alternative or diversified income streams for agriculture-dependent communities	Critics argue that the employment generated is often less than that generated by previous or alternative land uses; and that employment opportunities do not necessarily occur in the communities where plantations are established, with many of the jobs created being located some distance away. Concerns have also been raised about worker conditions in some countries
Local population	Provision of employment may result in new people shifting into communities, or retention of people who might otherwise have migrated elsewhere due to lack of local employment opportunities	Displacement of previous land uses by plantations may result in previous land users migrating elsewhere; if plantations provide fewer jobs, this may also be associated with decline in local population, and loss of local services or retail shops as a consequence
Water quality and quantity	Establishment of plantations can reduce soil erosion and improve water quality in catchments	Plantations may increase soil erosion during establishment and harvesting, and use of chemicals may result in water quality problems. Plantations may have high water use and reduce availability of water in local streams or aquifers
Biodiversity	Establishment of plantations can reduce logging pressure in native forests, and deforestation, and therefore support biodiversity. Plantations can also act as buffers for natural forest areas. Some types of plantations may also support particular flora and fauna that contribute to improved biodiversity outcomes in some regions	Plantations may reduce biodiversity if they replace natural ecosystems such as natural forests. The establishment of single species stands may reduce diversity compared to alternative uses of the same land. There may be increased invasive pests or diseases, or fires. Some concerns have also been expressed about whether some plantation tree species will become invasive weeds in nearby areas

Sources: Carrere and Lohmann (1996), Cossalter and Pye-Smith (2003), Schirmer (2007), McDermott (2012)

many other gaps in knowledge need to be filled; the three case studies we include exemplify the complex nature of achieving sustainability in the plantation sector, and the mix of business drivers, government incentives and community needs that must be considered; however, it must be recognised that there remains a lack of in-depth studies of both the business outcomes and sustainability implications of plantation expansion in many countries and contexts.

These case studies also highlight that debates about sustainability are not as simple as criticism of the practices of businesses: in all three, government incentives were an important driver of the way in which plantations were established, and hence of their sustainability outcomes. Globally, a majority of tree plantations have been established with the assistance of government subsidies or incentives such as government grants, tax incentives, low-interest loans, access to free or low-cost land, and provision of infrastructure or services (Bull *et al.* 2006). These incentives are intended to address perceived market failures and facilitate action that would otherwise happen more slowly, such as achieving national self-sufficiency in timber supply, supporting establishment of large-scale processing plants by guaranteeing them wood supplies, increasing economic opportunities in rural areas, and seeking to expand the environmental services provided by plantations, to name just a few. While designed to achieve positive outcomes, in practice many incentives can and have been abused or misused, or have resulted in unintended negative social, economic or environmental consequences. As a result, the provision of incentives is often associated with concerns about sustainability. The provision of incentives is argued to result in economic unsustainability through distorting market signals and thus encouraging establishment of timber plantations that are not economically viable (for example, establishment of plantations on low-productivity sites, use of inappropriate species, and establishment in locations with little or low access to markets) (Cossalter and Pye-Smith 2003). Incentives are also argued to encourage diversion of land use to plantations without consideration of whether other land uses would have greater economic, social or environmental benefit; and to lead to environmental unsustainability through creating incentives to establish plantations in environmentally sensitive areas, or using practices that are environmentally unsustainable (Bull *et al.* 2006), a criticism that applies particularly in our Indonesian case study.

On the other hand, incentives are often justified on the basis that they can provide opportunities to recognise benefits of plantations that can contribute to improved sustainability (for example, by paying for non-market services provided by plantations). In our third case study 60,000 smallholders in northeast and central Thailand are achieving livelihood benefits and supplying timber to a large plantation industry, while also providing a land use that is argued to reduce environmental problems such as soil erosion (Boulay *et al.* 2012). This was made possible initially through an incentive programme for smallholder plantations.

Case study 1: Large-scale conifer plantations in the United Kingdom (UK)

A desire to ensure self-sufficiency in timber production emerged in the UK after substantial clearance of forests led to shortages of timber during World War I and World War II. This drove development of government policies designed to encourage rapid expansion of tree plantations, including tax incentives for plantation establishment put in place from the 1970s. These incentives encouraged speculative investment in plantations by private investors seeking to reduce their tax burden. This investment created high demand for land, and increased the price of highly productive land. Some afforestation companies responded to lack of affordable land by draining peatland areas which had lower land values, and establishing plantations on the drained land. Following environmental protests in the 1980s, particularly around this practice being used in the Flow Country in northern Scotland, the tax incentives were removed by the government as of 1988. Debate continues about the economic viability of the plantations established on drained peatland, but there is reasonable consensus that at least some of the plantings caused ecological damage, and multiple projects since have sought to restore areas of peatland and associated habitat, with varying levels of success (sources: Anderson 2010; Warren 2000).

Case study 2: Small- and large-scale timber plantation programmes in Indonesia

Indonesia is one of the top ten pulp and paper producing countries globally (Obidzinski and Dermawan 2012). While pulp and paper production historically was dependent on harvesting of natural forest, a large and ambitious tree plantation programme has been established and has operated for more than two decades. Incentives to establish plantations were implemented in response to multiple factors, including declining harvest production from natural forests, and the imperative to provide timber supplies to the multiple large plywood, pulp and sawnwood processing facilities established in the country in recent decades. The plantation programme has two principle parts, not including the long-established teak and pine plantations in Java managed by the parastatal company Perum Perhutani. The first is the *Hutan Tanaman Industri* (HTI) policy, which commenced in 1990 and provided forest concessions to companies together with incentives to establish plantations on the concession land. By 2011, 249 companies had been issued permits covering 10 million ha of degraded natural forest, and around 4.9 million ha of plantations had been established (Wakker 2014). However, the HTI programme also led to clear-cutting of large areas of natural forest

(Barr *et al.* 2010, Pirard and Cossalter 2006), including on land that many argue has substantial ecological value, such as peatland areas in central Sumatra (Thorburn and Kull 2014). The volume of plantation production still falls well short of production capacity, meaning there is substantial incentive to supply processing facilities with illegally harvested logs from natural forest (Wakker 2014). Three major factors explain this outcome: the availability of cheap timber resources through forest conversion to supply mills means businesses profit more from this practice than more sustainable plantation establishment and management; there are many obstacles to the establishment and management of large-scale plantations in areas where claims to tenure or use rights on the land by local populations are the rule more than the exception, creating challenges for business viability (Wulan *et al.* 2004); and the Reforestation Fund which provided many of the incentives for plantation establishment has been erratically (mis)managed (Ernst & Young 1999; Wakker 2014). The programme has been associated with protests about lack of recognition of prior land users and their rights, and concern about social and economic impacts on local communities (Thorburn and Kull 2014). The second main policy aimed at expansion of plantations is the *Hutan Tanaman Rakyat* (HTR) programme, in which smallholder plantations are encouraged through the incentive of allocating smallholders rights to state forest land areas and providing seedlings and loans at subsidised interest rates. The HTR programme aimed to achieve planting of 5.4 million ha by 2016; by 2014, only 193,054 ha had permits issued for planting, and only 9,577 ha had been planted (data of Ministry of Forestry, September 2014), suggesting a substantial lack of business model viability, although few sustainability concerns have emerged. These disappointing results were explained by several implementation challenges (Obidzinski and Dermawan 2010): a slow and difficult process of identification of eligible land, high transportation costs because land made available was scattered, trade restrictions preventing transferring or inheriting permits, relative lack of economic attractiveness of allowed species compared with other commodities such as oil palm, and slow and complicated administrative processes.

Case study 3: Smallholder plantations in Thailand

In Thailand, large-scale plantation development in the 1980s was often associated with social conflict and protest, with local people protesting eviction of residents from areas established as plantations. These protests led, in 1989, to considerable tightening of how and what areas of land could be provided as concessions to large plantation companies (Barney 2004). In its place emerged widespread smallholder plantation development, with the government providing incentives in the form of free seedlings, fertiliser and 'soft' loans to encourage landholders to establish small plantation areas, and outgrower contracts with any of the multiple processing companies becoming common (Barney 2004). While the programme has had varying levels of plantation establishment over time, and most government incentives were removed during the Asian economic crisis of 1997 (Mahannop 2004), smallholder eucalypt plantation growing remains common, with a 2013 paper estimating that 60,000 smallholders were growing eucalypt plantations in Thailand (Boulay *et al.* 2013). Many smallholders choose to grow independently to supply a strong market for pulpwood in Thailand, while others grow under an 'outgrower' contract in which they grow trees and a company guarantees purchase of the product (Boulay *et al.* 2013). Boulay *et al.* (2012) found that smallholder plantations provided diversified income for some but not all of the farmers who grew them, but that some smallholders remained unable to benefit from the income diversification potential as they lacked ability to cope with loss of regular cash income during key growing years of the tree crop. Thus the smallholder programme has been successful, but potentially not equitable in terms of access for all landholders. The smallholder programme has established large areas of plantation – more than 300,000 ha by 2010 according to Boulay *et al.* (2012) – but without the social conflict that was a common feature of the previous highly contentious large-scale plantings driven by granting of large land concessions to companies. Some caution is needed, however: Boulay and Tacconi (2012) note the presence of negative perceptions of impacts of eucalypt plantations on the environment, and argue that the sustainability of the smallholder plantations has not been thoroughly examined.

Like many other aspects of plantation forests, government incentives provided to encourage plantation expansion should not be claimed to be either inherently good or bad for sustainability: their sustainability impacts will depend on how they are designed and implemented. Bull *et al.* (2006) argue that to avoid unintended negative outcomes governments should, amongst other things, create clearer links between instruments used to encourage plantation establishment, and sustainability outcomes. Explicitly considering all dimensions of sustainability and monitoring outcomes against these different dimensions will reduce the potential for unintended negative outcomes, while enabling positive sustainability outcomes. To do this, however, requires an understanding of the sustainability implications of plantations. The following three sections examine this in detail for three areas in which sustainability concerns commonly emerge: land ownership and tenure; deforestation and biodiversity; and employment and communities. We follow this with a discussion of how to enable positive outcomes in these core areas through designing plantations well, considering the issues of scale, and considering enabling conditions through the supply chain.

Land ownership and tenure

One of the largest business costs faced by plantation developers is the financial cost of accessing land on which to establish plantations. The drive to reduce costs of land can and has led to substantial tension between business and sustainability outcomes. Governments have often sought to encourage plantation expansion by offering cheap or free land to plantation companies. In many cases, this has been done with little consideration of the existing social, economic and environmental value of that land, leading to land being made available through the eviction of prior residents, through refusal to recognise claims of land ownership or tenure by indigenous groups or subsistence farmers (Chomitz 2007; Tauli-Corpuz and Tamang 2007), or through clearing natural forest or vegetation with important environmental values (discussed further in the next pages).

The widespread establishment of plantations on land without recognition of the rights of previous occupants or users of that land is one of the most commonly reported issues in conflicts over plantation establishment, including in two of our case study countries. A review by Gerber (2011) identified documented instances in which land users who lacked secure or recognised tenure were displaced from their land in order to make it available for plantation establishment, in countries as diverse as India, Thailand, Indonesia, South Africa, Cameroon and Chile. This occurred both on state-owned land under concession regimes, and in some cases on land sold to private investors.

While 'land grabs' for commercial plantation development are most commonly reported in countries with poorly documented or regulated tenure systems, concern about land tenure and land access rights extends beyond this. Even in countries with robust land tenure systems, concern has been expressed about the indirect effects of plantation expansion on land access for other landholders. For example, in the 1990s many farmers argued that plantation expansion in some rural areas of Australia drove up land prices and reduced their ability to compete with plantation companies to purchase land, thus reducing their ability to maintain a viable farm business (Schirmer and Tonts 2003). Meanwhile, in Ireland through the 1970s to 1980s farmers argued that land being allocated by the government for plantation establishment should have been distributed to farmers as part of longer-term processes of restoring land rights to Irish farmers after independence from Great Britain was achieved in the early 1900s (Schirmer 2007).

While cheap or free land has many business advantages, the first consideration when examining the sustainability of a plantation should be ensuring that the plantation is being established on appropriate land, under conditions of free, prior and informed consent. This requirement is now embedded in many voluntary certification schemes, such as the Forest Stewardship Council (FSC) and the Programme for the Endorsement of Forest Certification (PEFC) schemes, although different certification schemes vary in their interpretations of what constitutes free, prior and informed consent (Mahanty and McDermott 2013). The same certification schemes require consultation with stakeholders potentially affected by a plantation being established in their region. While addressing these issues is challenging, and may at first appear an imposition of costs that in itself reduces business viability of plantation companies, the cost of ensuring appropriate use of land is low compared to the social cost of plantation companies displacing previous land users without fair recognition, negotiation and compensation; and compared to the very high social and economic costs of the land use conflicts that often result when land is appropriated without free, prior and informed consent. Even considered from a purely business-oriented point of view, it is essential that businesses evaluate the true costs of conflicts, which can involve substantial delays in plantation establishment, damage to infrastructure and machinery, and increased management costs, amongst other consequences (Schirmer 2007).

Deforestation and biodiversity

The second common area in which sustainability opportunities and threats are reported for plantations is that of deforestation and biodiversity impacts, as is evident from Table 9.1.

Tree plantations are often promoted as a means of reducing deforestation by reducing pressure to log natural forests (see for example Sedjo 1999; Paquette and Messier 2010). On the other hand, they are argued by critics to contribute to deforestation and ecological degradation, particularly (although not only) when they are established on land cleared of natural forest or other important ecological communities (Carrere and Lohmann 1996; Cossalter and Pye-Smith 2003). This debate is exemplified by our second case study, in which a large-scale plantation programme is intended in part to provide wood supplies that act as an alternative to roundwood from natural forests, but which is argued to have itself contributed to deforestation through encouraging clearance of natural forest on forest concessions provided to companies as part of encouraging plantation establishment.

The idea that tree plantations can reduce deforestation is based on the idea that the timber they produce can substitute for timber from natural forests, and is supported by arguments relating to the relatively higher productivity of plantations in producing timber volume compared to most natural forests; and the availability of cleared or degraded land on which to establish plantations. A necessary condition is that new plantations are not established through clearing natural forests.

Despite the frequent promotion of plantations as having a role in reducing deforestation, few studies have evaluated whether this is in fact the case. Those that have use methods ranging from descriptive statistics to theoretical modelling and econometrics. Work by Dal Secco and Pirard (2015) suggests there is a convergence of results with evidence that the expansion of timber plantations is associated with reduced degradation of natural forests, but this does depend on the elasticity of demand, which determines whether plantation-grown timber will in fact replace demand for timber from natural forests, or whether there will simply be an overall increase in consumption of timber products. Similarly, the size of markets matters: the more open the economy the higher the risk that plantation supply will not substitute for natural forest production, but will instead add to it and increase supply to external markets (e.g. Pirard and Cossalter 2006). The effectiveness of plantations in substituting for natural forest timber also depends on scale: at the household level, several studies suggest success in replacing fuelwood sources used by households, for example (Dal Secco and Pirard 2015).

Additionally, tree plantation expansion might reduce natural forest degradation but increase deforestation at the same time, because it may contribute to lowering the economic value of natural forests where logging is no longer taking place, and create incentives for conversion of natural forests to other uses, particularly agriculture. This suggests a critical need for policies to accompany plantation expansion that address the potential for perverse outcomes, and for careful evaluation of the potential for such unintended effects. The lack of in-depth and robust evidence for the contention that tree plantations can reduce deforestation also emphasises a need for improved evidence to back the common claims made on this issue, whether the argument is that expansion of tree plantations reduces or enhances degradation of natural forests.

In the case of deforestation there is again a tension between business imperatives and sustainable development related to the cost of plantation management. In many cases, it is cheaper to access land for plantation establishment if that land has high ecological value but low economic value, rather than to access cleared land that has a higher price due to the higher value of its alternative land uses. In case study 1, a key driver for establishment of plantations on ecologically valuable peatland areas in Scotland was the high price of alternative sources of land that would not have led to the same environmental sustainability concerns, due to their viability for agricultural land uses. A common solution proposed to this is to better integrate smaller-scale plantings into existing landscapes, one that we examine in more detail later in the chapter.

Employment and communities

Many of the social and economic sustainability concerns raised about plantations centre on their effects on rural community life. As McDermott (2012) identifies, critics argue that plantations displace previous residents and land uses, reduce employment, and lead to loss of population and reduced economic viability. Proponents argue that plantations provide new employment and income diversification opportunities, and by doing so support rural population levels and hence local communities. As with many other issues related to sustainability of plantations, there is limited evidence to decide the ongoing debates about the employment and community effects of plantations (McDermott 2012).

Despite being an apparently simple question, identifying the true 'impact' of plantations on employment and local communities requires detailed analysis of various factors, including: the amount of employment generated by the plantation compared to previous land uses; the spatial location of the jobs, particularly whether people who work on the plantation live in local communities or some distance away; the wages paid to plantation workers compared to those paid to people for alternative land uses; and the working conditions provided for workers.

Detailed studies in Australia suggest that changing land use from agriculture to plantations can result in either a net gain or a net loss of employment, depending on what type of agricultural land use is replaced and the type of plantation being established; and that changing land use to plantations is associated with complex spatial changes in the location of jobs, which means there are employment benefits for some communities and losses for others (Williams and Schirmer 2012). For example, long-rotation plantations grown for sawnwood production produce higher rates of employment per hectare compared to short-rotation plantations grown for pulpwood production. The same study highlighted that over time, increasing mechanisation of most land uses has reduced the employment they generate, including agricultural land uses as well as plantations (Williams and Schirmer 2012). This suggests that increasing mechanisation in plantations, a change that often improves business efficiency and viability, but is frequently associated with decreasing labour requirements (Bayne and Parker 2012) should not be argued to reduce economic benefits of plantations unless it is occurring in the absence of similar mechanisation of alternative land uses.

As with other areas of sustainability, whether a plantation has positive, neutral or negative outcomes for employment depends on the way that the plantation and the supply chain that surrounds it is designed. For example, Pirard and Mayer (2008) argue that successfully integrating plantations in the landscape can enable positive outcomes by creating opportunities for workers to combine work at the plantation with associated cash wages from other activities, such as taking care of cattle or gardens.

Designing sustainable plantations

The widespread debate about sustainability of plantations highlights the importance of carefully designing plantations. Design here means not just the configuration, location and species planted, but the design of governance systems, markets and supply chains to support sustainable practice in the plantation sector. Design is critical: asking plantation businesses to bear the burden of additional costs to ensure sustainability in the absence of a broader system of governance and markets that requires and rewards sustainable practice will likely lead to business failure as, in the absence of good governance and market support structures, cheaper unsustainably produced plantation roundwood will outcompete more sustainable product.

Throughout this chapter, we emphasise that plantation forestry cannot be characterised as inherently 'sustainable' or 'unsustainable'. Rather, each situation in which plantations are being established should be evaluated to identify potential sustainability challenges and opportunities. These must then be addressed and enacted throughout the life cycle of plantation forestry business activities, from proposal and planning to harvest and replanting. This can be supported by business systems and processes that ensure such evaluation and action is in-built as part of plantation design, implementation and management. These business systems must go beyond checklists of different concerns or issues that have emerged with previous plantations: a new plantation may present new or unique sustainability considerations, and businesses need to have processes that proactively identify not just the issues that have arisen in the past, but also emerging challenges, thus enabling these to be addressed before they become a sustainability problem. A key challenge in achieving this is a dearth of evidence on which to make good decisions about the sustainability of plantations: much of the debate about sustainability is assertion- rather than evidence-based. Spatially and temporally explicit work on sustainability of plantations in different situations is a critical need, with examples such as the work of Baral *et al.* (2014) providing an exemplar of the type of evidence necessary to inform stakeholders and decision processes.

While a growing literature focuses on when and why sustainability problems may emerge from plantations, particularly industrial plantations, very little work has examined what conditions enable sustainable plantations to be established by businesses (Kröger 2013). There is, however, a broad body of work that suggests four core areas for consideration in the design of sustainable plantations. First, the scale of plantation establishment should be considered. Second, the role of the supply chain must be considered. Third, dialogue and engagement with stakeholders is essential. Fourth, good governance systems – which can encompass and drive many of the first three areas – are essential, whether they be government-driven, or voluntary through certification systems such as the FSC or PEFC. Each of these four areas is considered below.

Enabling sustainability through scale: Is small-scale the answer?

Worldwide, many of the sustainability concerns expressed about plantations are focused on industrial plantations, specifically those plantations established in large single areas, usually using monocultural stands, owned by large national or multinational corporations, and producing pulpwood or biofuel (Schirmer 2007; Gerber 2011; Kröger 2013):

> there is a set of 'specific grievances' typically arising against corporate pulpwood plantations: pollution of soil and waters in the investment

> area; expansion to traditional communities' lands; rural mechanization and ensuing unemployment; industrial pollution; increased traffic due to logistical operations; outsourcing and the degradation of working conditions; creation of food insecurity by monocultures … A universally applied investment model, such as the large-scale pulp model, tends to create a set of broadly similar industry-specific grievances across different contexts.
>
> (Kröger 2013: 30)

Gerber (2011) argues that the reason for higher rates of conflict over large-scale industrial plantations is because the scale of impacts of plantations corresponds to their size, with large-scale plantations more likely to be the subject of conflict due to their greater social, economic and environmental impacts.

Given that the types of sustainability concerns documented in Table 9.2 are most commonly raised in relation to large-scale plantations, are small-scale plantations the answer to sustainability concerns? Small-scale plantations – typically a few hectares established by individual landholders who sell the timber to contribute to their household income – are argued by many to avoid some or all of the problems associated with large-scale plantations, such as displacement of residents, lack of provision of local employment and environmental impacts. Work by Schirmer (2007) and Gerber (2011) highlights that, to date, little conflict has emerged over small-scale, diverse tree plantations, even in regions where the total area of these small plantations was relatively large. Small-scale forestry is considered more socially acceptable than large-scale plantations, particularly because it is argued to provide income-earning benefits to the landholders who adopt it, to provide employment opportunities to locals rather than to external 'outsider' corporations, and to address various environmental degradation problems as well as provide benefits such as shade and shelter to livestock. The experiences of countries such as Thailand (described in case study 3) and Vietnam support these broader findings: in these countries, large numbers of smallholder plantations have been established, and despite their aggregate area being relatively large, have attracted less criticism than industrial plantations – although it should be noted they are not criticism-free, with some concerns being raised about equity of access to smallholder trees, with richer households more likely to be able to access them (McElwee 2009; Boulay *et al.* 2012; Sikor 2012).

Despite these encouraging examples, in most countries small-scale plantations, such as agroforestry, farm forestry, and other smallholder tree plots, make up a relatively small proportion of the total tree plantation estate, and many agroforestry programmes achieve lower than hoped for rates of adoption (Glover *et al.* 2013). This failure to achieve desired rates of adoption is explained in many ways in the literature, usually with reference to the many factors that reduce landholder willingness to take on small-scale forestry, such as concerns about taking on a long-term investment; low financial returns from agroforestry; reductions in land use flexibility associated with growing a long-term crop such as trees; and a view that growing trees on farmland is unacceptable, amongst others (see e.g. Schirmer *et al.* 2000; Pattanayak *et al.* 2003; Mercer 2004). For example, in Australia, Schirmer *et al.* (2008) identified that landholders who were given the option of growing eucalypts for commercial wood production on their land with a guaranteed purchaser of the pulpwood at the end of the rotation, versus selling their land to a plantation company, typically chose the latter option. They did this as selling land had greater economic benefits for them, and enabled them to purchase new land that better fitted their farming priorities, whereas leasing land required them to commit their land long-term to a land use they couldn't change.

This suggests that small-scale plantations often have important business viability challenges that reduce uptake by the landholders who are essential to their widespread adoption, despite their widespread promotion in the literature. It is essential to overcome the problem of creating viable business opportunities in the small-scale sector if small-scale plantations are to become the solution to plantation sustainability issues many argue they could be.

Business imperatives are also argued to drive the focus on establishment of large-scale plantations in preference to small-scale plantations. Large-scale plantations are argued to provide greater economies of scale, reducing the cost per unit of production and improving competitiveness in the global timber market (Schirmer 2007). However, this economic imperative is debated, and in some cases supply chain relationships have been developed that enable the aggregation of small-scale growers to supply large markets (see for example case study 3, and Nawir and Santoso 2005), thus enabling economies of scale without having to establish large single areas of plantation.

Enabling sustainability through the supply chain

Sustainability considerations about plantations are not limited to the site of the plantation itself. It is also critical to consider sustainability issues in the broader supply chain that plantation forestry forms part of, and how the supply chain can support sustainable practice. This means that plantation companies need to consider not only the sustainability of their own practices, but of their suppliers and of the processing businesses and markets they sell products into. Is the plantation supplying a processing plant about which there are sustainability concerns such as poor working conditions, pollution, corruption or other issues? Is the plantation located in an area with access to a viable market that enables the realisation of the income earning opportunities promised by plantation developers? A plantation is not sustainable unless it is part of a supply chain in which there are appropriate social, economic and environmental outcomes through to the final market destinations of products.

Although supply chain analysis is now widely recognised as critical to ensuring sustainability and corporate social responsibility (see for example Andersen and Skjoett-Larsen 2009), full supply chain analysis of sustainability remains an emerging area in the field of plantation forestry (see for example Zuo *et al.* 2009). Most voluntary certification schemes have some supply chain focus as part of certifying plantations, the extent of which varies depending on the type of certification involved.

Businesses are increasingly recognising that to be viewed as legitimate by the broader public, and hence have a 'licence' of social acceptability that enables them to operate without protest or conflict, they must consider the sustainability of the entire supply chain they are part of, rather than just of their own operations (Vidal *et al.* 2010). Perhaps the most active consideration of the broader supply chain can be seen in the activism of environmental non-government organisations (ENGOs), who target the entire supply chain as part of market campaigns to achieve change in what they view as unsustainable practices. This emphasises that unsustainability at one point in the supply chain has far reaching implications, through to consumer markets in which the goods produced through unsustainable practices are sold (Schirmer 2013). Plantation businesses must consider sustainability in their broader supply chain, as protests against unsustainable practices in any part of that chain can have very real effects on their own business viability, as activists target the supply chain in order to achieve change. What may at first seem like a practice in other parts of the supply chain that is disconnected from the day–to-day business priorities of a plantation company in fact presents a substantial threat to their market success, and must be actively addressed by that company as part of ensuring both sustainable practice *and* their own business viability.

Enabling sustainability through dialogue

Sustainable practice requires engagement between stakeholder groups at a number of different levels: both the communities in which plantations are established, and other stakeholders who have an interest in plantation forestry. Kröger (2013), in one of the few studies to examine why conflict had *not* arisen over a large-scale plantation development, identified that a critical factor in the success of the Suzano pulp plantation investment in the Eastern Amazon was the presence of good relationships between the company involved and local communities. This finding echoes that of many other studies that emphasise the importance of having thorough and meaningful public participation processes, both to prevent conflict over plantation practices (see for example Gordon *et al.* 2013; Dare *et al.* 2014), and to resolve these conflicts if they do emerge (Dhiaulhaq *et al.* 2014). The Forests Dialogue[1] and the New Generations Plantations Platform[2] are two examples of dialogue processes that connect plantation stakeholders at local, national and international levels.

The need for good processes of public participation – variously labelled consultation, collaboration, community engagement and many other names, each of which can be defined in multiple ways – is well recognised. However, to be successful in ensuring sustainable practice this public participation must be meaningful. 'Meaningful' means that there must be an ability to modify or change plantation practices in response to the issues, questions and opportunities identified during public participation processes (Schirmer 2007; Dare *et al.* 2010). Tokenistic consultation processes, in which plantation companies invite people to submit their views but fail to act on or respond to them, are likely to worsen relationships between plantation businesses and the communities and stakeholders with an interest in their activities; and reduce the likelihood that potential threats to sustainable practice will be identified and acted on.

Many methods can be used to ensure appropriate communication and collaboration with communities and stakeholders (see for example Dare *et al.* 2010). These should be tailored to the local cultures and practices in different parts of the globe in which plantations are established. While often dismissed as cost prohibitive, the costs of proactive communication must be weighed against the costs of delays in plantation operations resulting from protest or conflict over plantations (Dare *et al.* 2010). In many cases, the costs of delays or lost production are likely larger than the investment needed to improve communication; the evidence of widespread conflict over plantations worldwide suggests that conflict-related expenses represent one of the more substantial costs faced by plantation companies in many regions (Mola-Yudego and Gritten 2010).

Enabling sustainability through effective governance

Sound, effective and fair governance systems are essential to ensuring the sustainability of plantations. For example, ensuring plantations do not displace previous land uses or users requires the presence of just and effective land tenure systems that protect the rights of both the environment and people. Good governance may be driven by government or by voluntary governance systems such as the PEFC or FSC certification schemes already prominent in the forestry sector (or, ideally, involve both). Given that good governance is implemented at multiple scales and in multiple ways, Kanowski (2010) recommends four steps to implementing good governance for the forest sector: (i) establishing the evidence base needed to support good governance, and ensure policy makers have access to this information; (ii) agreeing on the contributions plantations should make – requiring strong processes of stakeholder participation and dialogue to achieve consensus where possible on what is a 'sustainable' plantation; (iii) designing governance regimes that address the issues and needs identified in the first two steps; and (iv) ensuring forest practice systems, including monitoring and evaluation systems, are established to support and meet the requirements of the governance regimes in place. Achieving these four steps requires implementing many of the areas of practice discussed earlier in this chapter, particularly establishing a solid evidence base regarding sustainability issues that are often contested, and developing systems of constructive two-way dialogue.

Future directions

For their business to succeed, plantation forest managers must understand that the markets for the goods they produce depend on their social licence to operate, in other words on whether consumers and stakeholders believe their business is sustainable enough to be acceptable. Achieving social licence in turn depends on the sustainability of plantations, and public perceptions of that sustainability. Plantation businesses are increasingly recognising this.

Perhaps the highest profile example of a public commitment to changing practices in order to achieve social licence to operate is the announcement in 2013 by Asia Pulp and Paper (APP) that they would implement substantial changes to their practices, including ending their clearing of natural forests and guaranteeing free, prior and informed consent. This decision was made after a sustained campaign criticising APP's practices, particularly in Indonesian forests (Dieterich and Auld in press), which substantially reduced their social licence through consequences such as the Forest Stewardship Council disassociating APP and subsidiary companies and terminating certifications held by those companies. This is one of many examples in which plantation companies are committing to changed practices, although there is debate both about the 'genuineness' of the commitments made, and whether highly public pressure campaigns are effective in achieving change to sustainable management in the longer term (Dieterich and Auld in press). Despite these caveats, and in the context of business and sustainability, plantation businesses must recognise that social licence to operate is dependent on public perceptions of the sustainability of their practices, and proactively ensure and communicate sustainable practice.

The large-scale monoculture plantations preferred by many plantation growers for their perceived business advantages, and which have dominated expansion of plantation forestry in recent decades, are also the plantations most associated with sustainability concerns. The small-scale planted forests promoted by many as the solution to the concerns associated with large-scale corporate-owned plantations have in most countries not achieved the hoped-for widespread adoption by landholders, and have also attracted some criticism. Moving forward, a more sophisticated discussion about the benefits and costs of tree planting at different scales is needed, together with design of programmes that combine the best of both, while minimising the potential negative outcomes of both small- and large-scale plantings.

Plantation forests are often argued to provide benefits on the one hand and costs on the other, often for the same issue, such as employment, biodiversity or water quality. On the one hand, they are promoted as having potential to reduce deforestation through providing an alternative wood supply; on the other, they are criticised as causing deforestation, principally when they are established on land cleared of natural forest for the purpose of establishing the plantation. They are promoted as a potential generator of jobs and income in rural communities, but concerns are also raised about whether plantations generate the same number of jobs as other land uses, and whether the jobs and income generated go to local communities or to outsiders.

How can the sector reconcile the business opportunities presented by plantations with the widespread concerns about sustainability? First, it must take seriously the real concerns about sustainability raised by critics. These are reasonably widely documented, but more systematic documentation of sustainability concerns is needed, together with evaluation of the evidence where there are competing or conflicting claims. Second, better understanding is needed of when and why plantations work *well*. It is dangerous to focus only on identifying problems, as this fails to recognise what conditions enable the establishment of successful plantations both in terms of providing economic return, and in terms of broader social, environmental and economic sustainability. Third, those who promote simple solutions to complex problems – such as the use of small-scale plantations as a panacea for the problems argued to result from industrial plantations – must develop more complex, nuanced and realistic analyses that recognise the barriers that prevent more widespread use of some of these proposed 'solutions' to concerns over sustainability of plantations. Mature debate about the consequences of plantations is needed that demands better evidence for claims about both the promises and the perils of plantation forestry if there is to be a shift to truly sustainable practice in the plantation forest sector. Finally, achieving sustainability in the plantation sector depends on what is happening in the rest of the global forest sector. A key challenge is the price signals in the market: where plantation-grown roundwood competes with wood harvested in natural forests that are unsustainably managed, prices often do not reflect the true cost of producing sustainably grown wood. Achieving the price signals needed for both viable plantation businesses and sustainably grown plantation roundwood requires reform not just in the plantation sector, but in the broader global forest sector.

Notes

1 http://theforestsdialogue.org/initiatives/IMPF.
2 http://newgenerationplantations.org.

References

Andersen, M. and Skjoett-Larsen, T. (2009). Corporate social responsibility in global supply chains. *Supply Chain Management: An International Journal*, 14(2), 75–86.

Anderson, R. (2010). *Restoring afforested peat bogs: results of current research*. Forestry Commission Research Note, Forestry Commission, Roslin, UK.

Arevalo, J., Ochieng, R., Mola-Yudego, B. and Gritten, D. (2014). Understanding bioenergy conflicts: case of a Jatropha project in Kenya's Tana Delta. *Land Use Policy*, 41, 138–148.

Baral, H., Keenan, R.J, Sharma, S., Stork, N.E. and Kasel, S. (2014). Economic evaluation of ecosystem goods and services under different landscape management scenarios. *Land Use Policy*, 39, 54–64.

Barney, K. (2004). Re-encountering resistance: plantation activism and smallholder production in Thailand and Sarawak, Malaysia. *Asia Pacific Viewpoint*, 45(3), 325–339.

Barr, C., Dermawan, A., Purnomo, H. and Komarudin, H. (2010). *Financial governance and Indonesia's Reforestation Fund during the Soeherto and post-Soeharto periods, 1989–2009: A political economy analysis of lessons for REDD+*, Occasional Paper 52, Center for International Forestry Research, Indonesia.

Barua, S.K., Lehtonen, P. and Pahkasalo, T. (2014). Plantation vision: potentials, challenges and policy options for global industrial forest plantation development. *International Forestry Review*, 16(2), 117–127.

Bayne, K.M. and Parker, R.J. (2012). The introduction of robotics for New Zealand forestry operations: forest sector employee perceptions and implications. *Technology in Society*, 34, 138–148.

Boulay, A. and Tacconi, L. (2012). The drivers of contract eucalypt farming in Thailand. *International Forestry Review*, 14(1), 1–12.

Boulay, A., Tacconi, L. and Kanowski, P. (2012). Drivers of adoption of eucalypt tree farming by smallholders in Thailand. *Agroforestry Systems*, 84: 79–189.

Boulay, A., Tacconi, L. and Kanowski, P. (2013). Financial performance of contract tree farming for smallholders: the case of contract eucalypt tree farming in Thailand. *Small-scale Forestry*, 12(2), 165–180.

Bull, G.Q., Bazett, M., Schwab, O., Nilsson, S., White, A. and Maginnis, S. (2006). Industrial forest plantation subsidies: impacts and implications. *Forest Policy and Economics*, 9(1), 13–31.

Carle, J. and Holmgren, P. (2008). Wood from planted forests: a global outlook 2005–2030. *Forest Products Journal*, 58(12), 6–18.

Carrere, R. and Lohmann, L. (1996). *Pulping the South. Industrial Tree Plantations and the World*. Zed Books, London.

Chomitz, K.M. (2007). *At Loggerheads? Agricultural Expansion, Poverty Reduction, and Environment in the Tropical Forests*. World Bank, Washington D.C.

Cossalter, C. and Pye-Smith, C. (2003). *Fast-wood Forestry: Myths and Realities*. Vol. 1. CIFOR.

Dal Secco, L. and Pirard, R. (2015). Do tree plantations support forest conservation? CIFOR Infobrief No. 110, Center for International Forestry Research, Indonesia.

Dare, M., Schirmer, J. and Vanclay, F. (2010). *Handbook for Operational Community Engagement within Australian Plantation Management. Improving the Theory and Practice of Community Engagement in Australian Forest Management*. Cooperative Research Centre for Forestry, Hobart, Tasmania.

Dare, M., Schirmer, J. and Vanclay, F. (2014). Community engagement and social licence to operate. *Impact Assessment and Project Appraisal*, 32(3), 188–197.

Del Lungo, A., Ball, J. and Carle, J. (2006). *Global planted forests thematic study. Results and analysis*. Planted Forests and Trees Working Papers, FAO.

Dhiaulhaq, A., Gritten, D. and De Bruyn, T. (2014). *Mediating forest conflicts in Southeast Asia*. RECOFTC Issue Paper No. 2. ECOFTC – The Center for People and Forests, Bangkok, Thailand.

Diaz, D., Hamilton, K., Johnson, E., Kandy, D. and Peters-Stanley, M. (2011). *State of the Forest Carbon Markets 2011: From Canopy to Currency*. Ecosystem Marketplace, Washington, DC.

Dieterich, U. and Auld, G. (in press). Moving beyond commitments: creating durable change through the implementation of Asia Pulp and Paper's forest conservation policy. *Journal of Cleaner Production*.

Ernst & Young (1999). Special audit of the Reforestation Fund. Final report, Jakarta.

FAO (2010). *Global Forest Resources Assessment 2010 Main Report*. FAO.

Gerber, J.F. (2011). Conflicts over industrial tree plantations in the south: who, how and why? *Global Environmental Change*. 21(1), 165–176.

Glover, E.K., Ahmed, H.B. and Glover, M.K. (2013). Analysis of socio-economic conditions influencing adoption of agroforestry practices. *International Journal of Agriculture and Forestry*, 3(4), 178–184.

Gordon, M., Schirmer, J., Lockwood, M., Vanclay, F. and Hanson, D. (2013). Being good neighbours: current practices, barriers, and opportunities for community engagement in Australian plantation forestry. *Land Use Policy*, 34, 62–71.

Government of Vietnam (2011). Report on the implementation of the *five* million ha new afforestation project and the forest protection and development program in the 2011–2020 period, Hanoi, Vietnam.

Jurgensen, C., Kollert, W. and Lebedys, A. (2014). *Assessment of industrial roundwood production from planted forests*. Working Paper FP/48/E, FAO Planted Forests and Trees Working Paper Series, Food and Agriculture Organization of the United Nations, Rome.

Kanowski, P.J. (2010). Policies to enhance the provision of ecosystem goods and services from plantation forests. Chapter 7 in Bauhus, J.*et al.* (Eds). *Ecosystem Goods and Services from Planted Forests*. Earthscan, Oxford, 171–204.

Kröger, M. (2013). Grievances, agency and the absence of conflict: the new Suzano pulp investment in the Eastern Amazon. *Forest Policy and Economics*, 33: 28–35.

Mahannop, N. (2004). The development of forest plantations in Thailand. In Enters, T. and Durst, P.B. *What Does It Take? The Role of Incentives in Forest Plantation Development in Asia and the Pacific*. FAO, Office for Asia and the Pacific, 211–236.

Mahanty, S. and McDermott, C.L. (2013). How does 'Free, Prior and Informed Consent' (FPIC) impact social equity? Lessons from mining and forestry and their implications for REDD+. *Land Use Policy*, 35, 406–416.

McDermott, C.L. (2012). *Plantations and Communities: Key Controversies and Trends in Certification Standards*. Forest Stewardship Council.

McElwee, P. (2009). Reforesting 'bare hills' in Vietnam: social and environmental consequences of the 5 million hectares reforestation program. *Ambio*, 38(6), 325–333.

Mercer, D.E. (2004). Adoption of agroforestry innovations in the tropics: a review. *Agroforestry Systems*, 61(1–3), 311–328.

Mola-Yudego, B. and Gritten, D. (2010). Determining forest conflict hotspots according to academic and environmental groups. *Forest Policy and Economics*, 12(8), 575–580.

Nawir, A.A. and Santoso, L. (2005). Mutually beneficial company–community partnerships in plantation development: emerging lessons from Indonesia. *International Forestry Review*, 7(3), 177–192.

Obidzinski, K. and Dermawan, A. (2010). Smallholder timber plantation development in Indonesia: what is preventing progress? *International Forestry Review*, 12(4), 339–348.

Obidzinski, K. and Dermawan, A. (2012). Pulp industry and environment in Indonesia: is there sustainable future? *Regional Environmental Change*, 12(4), 961–966.

Paquette, A. and Messier, C. (2010). The role of plantations in managing the world's forests in the Anthropocene. *Frontiers in Ecology and the Environment*, 8: 27–34.

Pattanayak, S.K., Mercer, D.E., Sills, E. and Yang, J.C. (2003). Taking stock of agroforestry adoption studies. *Agroforestry Systems*, 57(3), 173–186.

Pin Koh, L. and Ghazoul, J. (2008). Biofuels, biodiversity, and people: understanding the conflicts and finding opportunities. *Biological Conservation*, 141, 2450–2460.

Pirard, R. and Cossalter, C. (2006). *The revival of industrial forest plantations in Indonesia's Kalimantan provinces. Will they help eliminate fiber shortfalls at Sumatran pulp mills or feed the China market?* CIFOR Working Paper No. 37, Center for International Forestry Research, Indonesia.

Pirard, R. and Mayer, J. (2008). Complementary labor opportunities in Indonesian pulp plantations, and implications for land use. *Agroforestry Systems*, 76(2), 499–511.

Righelato, R. and Spracklen, D. (2007). Carbon mitigation by biofuels or by saving and restoring forests. *Science*, 317, 902.

Rudel, T.K. (2009). Tree farms: driving forces and regional patterns in the global expansion of forest plantations. *Land Use Policy*, 26(3), 545–550.

Schirmer, J. (2007). Plantations and social conflict: exploring the differences between small-scale and large-scale plantation forestry. *Small-scale Forestry*, 6(1), 19–33.

Schirmer, J. (2013). Environmental activism and the global forest sector. In Hansen, H., Panwar. R. andVlosky, R. (Eds). *The Global Forest Sector: Changes, Practices, and Prospects*. CRC Press, Boca Raton, 203–235.

Schirmer, J. and Tonts, M. (2003). Plantations and sustainable rural communities. *Australian Forestry*, 66(1), 67–74.

Schirmer, J., Kanowski, P. and Race, D. (2000). Factors affecting adoption of plantation forestry on farms: implications for farm forestry development in Australia. *Australian Forestry*, 63(1), 44–51.

Schirmer, J., Loxton, E. and Campbell-Wilson, A. (2008). *Impacts of land-use change to farm forestry and plantation forestry: a survey of landholders*. Cooperative Research Centre for Sustainable Production Forestry, Hobart.

Sedjo, R.A. (1999). The potential of high-yield plantation forestry for meeting timber needs. In *Planted Forests: Contributions to the Quest for Sustainable Societies*. Springer, Netherlands, 339–359.

Sikor, T. (2012). Tree plantations, politics of possession and the absence of land grabs in Vietnam. *Journal of Peasant Studies*, 39(3–4), 1077–1101.

Tauli-Corpuz, V. and Tamang, P. (2007). *Oil palm and other commercial tree plantations, monocropping: impacts on indigenous peoples' land tenure and resource management systems and livelihoods*. In UN Permanent Forum on Indigenous Issues Working Paper, E/C (Vol. 19).

Thorburn, C.C. and Kull, C.A. (2014). Peatlands and plantations in Sumatra, Indonesia: complex realities for resource governance, rural development and climate change mitigation. *Asia Pacific Viewpoint*, 56, 153–168.

Vidal, N.G., Bull, G.Q. and Kozak, R.A. (2010). Diffusion of corporate responsibility practices to companies: the experience of the forest sector. *Journal of Business Ethics*, 94(4), 553–567.

Wakker, E. (2014). *Indonesia: Illegalities in Forest Clearance for Large-Scale Commercial Plantations*. Forest Trends, Washington, DC, and Aidenvironment Asia, Amsterdam. www.forest-trends.org/documents/files/doc_4528.pdf.

Warman, R.D. (2014). Global wood production from natural forests has peaked. *Biodiversity and Conservation*, 23, 1063–1078.

Warren, C. (2000). 'Birds, bogs and forestry' revisited: the significance of the flow country controversy. *The Scottish Geographical Magazine*, 116(4), 315–337.

Williams, K.J. and Schirmer, J. (2012). Understanding the relationship between social change and its impacts: the experience of rural land use change in south-eastern Australia. *Journal of Rural Studies*, 28(4), 538–548.

World Commission on Environment and Development (1987). *Our Common Future* (Vol. 383). Oxford University Press, Oxford.

Wulan, C.Y., Yasmi, Y., Purba, C. and Wollenberg, E. (2004). *Analisa konflik sector kehutanan di Indonesia 1997–2003*. Center for International Forestry Research, Indonesia.

Xu, J. (2011). China's new forests aren't as green as they seem. *Nature*, 477, 371.

Zuo, K., Potangaroa, R., Wilkinson, S. and Rotimi, J.O. (2009). A project management prospective in achieving a sustainable supply chain for timber procurement in Banda Aceh, Indonesia. *International Journal of Managing Projects in Business*, 2(3), 386–400.

10 The emerging bio-economy and the forest sector

Anders Roos and Matti Stendahl

The bio-economy – meaning and background

The concept of the bio-economy, often used interchangeably with "bio-based economy" or "the Green Economy", has become popular in recent years and the transition of the global economy to a bio-economy has been a central topic in academic and policy realms. The forest sector is expected by many analysts to take center stage in this process. In this chapter we discuss the potential for the bio-economy to, as it claims, join economic growth with sustainability, and the role of the forest sector in advancing this process. The key question is "if" – or under which conditions – the forest sector can truly propel a revitalized world economy that is based on the sustainable use of bio/forest-resources. Does the vision of the forest sector providing bio-based products align with the vision of an efficient, competitive and sustainable forest sector and forest industry processes – the fundamentals of a bio-economy (Figure 10.1)? To what extent can forests and the related industry provide forest-based goods and services that not only fulfill the ecological and social requirements in their management and processes, but also provide superior products, services, user value and competitiveness in the market?

Examination of the contribution of the forest sector to the bio-economy warrants a closer look at the latter concept. As a relatively recent concept, the bio-economy notion stems from advances in biosciences (such as genomics, bioengineering and biorefinery processing) that occurred from the 1990s. These advances were based on mapping, designing and building genetic material and new substances that pervaded many industry sectors, including chemicals, cosmetics, pharmaceuticals, foodstuffs, energy and computers (Enriquez-Cabot 1998). The bio-economy became envisioned as an avenue for the agricultural sector to contribute "to a more sustainable and profitable balance of production and markets" (Hardy 2002, 15). With growing concerns for sustainability, the bio-economy concept has since expanded into the policy sphere, where it has widened in scope to signify economic growth that is decoupled from an unsustainable use of fossil resources, featuring in policy documents as a way towards sustainable economic progress in the

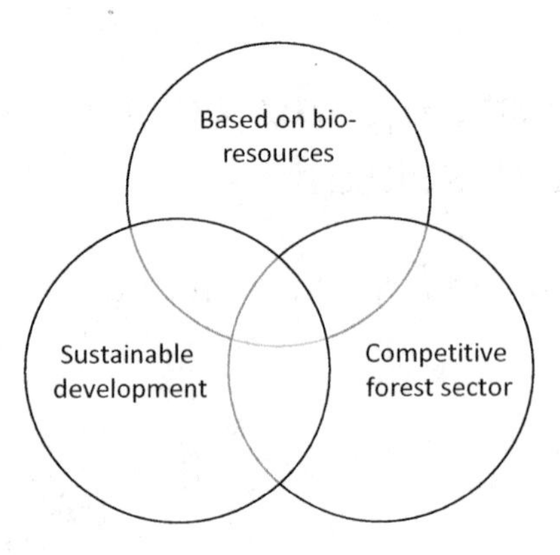

Figure 10.1 The bio-economy and the forest sector

world economy. The Organisation for Economic Co-operation and Development (OECD) noted the twin challenges of providing goods and services to an increasing, more demanding world population with increasing purchasing power – while maintaining the capacity of natural capital, resource stocks, land and ecosystems (OECD 2011, 9). It emphasized the economic potential of the bio-economy to "open up new sources of growth" that could not only spur innovation and investment, but also be consistent with resilience of ecosystems (OECD 2011, 9–11). The bio-economy has slightly diverse meanings in different documents, ranging from the narrow definition of exclusively advanced biotechnologies, to encompassing almost all economic uses of bio-based resources. This difference among users of the concept and "drift" over time in definition of the bio-economy has resulted in a somewhat ambiguous discussion around the concept in both political and scientific discourse (Kleinschmit *et al.* 2014). The US White House Blueprint on the bio-economy gave a restricted definition of the concept, alluding to growth prospects of the bio-economy, primarily based on advanced bio-technologies: "Economic activity that is fueled by research and innovation in the biological sciences" (White House 2012, 1). The bio-economy strategy of the European Commission, envisions the bio-economy more broadly: including also the traditional forest and agricultural sectors contributing to an "innovative, resource efficient and competitive society that reconciles food security with the sustainable use of renewable resources for industrial purposes, while ensuring environmental protection". Still, the common goal in most documents is sustainable growth and employment (EC 2012, 2). These views can also be compared with the UN Rio+20 document that describes the green economy as a means toward poverty alleviation and sustainable global development (United Nations 2012, 2). This range of definitions suggests that the bio-economy concept emerges as a "mixed sources" discourse that borrows arguments from different world views, such as the technocratic/instrumental approach, and the ecological modernization and sustainable development discourses (Pülzl *et al.* 2014).

As growth rates in western countries have remained low or even negative in the early years of the twenty-first century, the bio-economy symbolized a new direction for revitalized economies sought by businesses and policy makers – a direction that incorporates increasing concerns for climate and the environment together with a reduced strain on natural resources. Economic policy documents underscore that advances should materialize in a competitive economy through innovation and entrepreneurship. Thus, the bio-economy concept has influenced universities, enterprises and funding organizations to invest resources in developing new uses of the bio-resource. The aspired-to future potential of the bio-economy is now reflected in supranational and national research and innovation efforts, based on bio-resources, aiming for new superior technologies and finally product offerings on the market (Staffas *et al.* 2013).

The role of forests in the bio-economy was outlined by Duchesne and Wetzel (2003), foreseeing an increased forest contribution in a growing range of biological products and the intensified use of biotechnology in biomass processing. Applying a broad definition of a bio-economy, the United Nations Economic Commission for Europe/Food and Agriculture Organization of the United Nations (UNECE/FAO) defined different aspects where the forest sector contributes in the transition to a green economy: through sustainable production and consumption of forest products, the low-carbon forest sector, job creation and forest ecosystem services (UNECE/FAO 2011, 1). Forest sector associations are also branding themselves as key actors in the transition and as providers of economic development, jobs, innovation and services (Swedish Forest Industries Federation 2012; Puddister *et al.* 2011). Industry representatives in, for example, Canada, Sweden and Finland, refer to the sector's record of increasingly sustainable logging methods, and production of both established and new bio-products. The forest sector may also contribute to a bio-economy via its role in providing livelihoods in developing countries through products and ecological services (UNEP 2011b, 21).

The bio-economy idea has met criticism as well: sceptics of the concept claim that it is designed as a license for forest corporations to continue harmful practices and also develop new risky technologies, such as genetically engineered trees (Hall *et al.* 2012). Hidden conflicts and contradictions between the industry and other stakeholders, such as farmers and environmental NGOs were identified by Richardson (2012). Pülzl *et al.* (2014) stated that the bio-economy discourse in policy and corporate documents is, to a large extent, based on economic arguments whereas global governance and civic participation considerations are virtually neglected, and problematic areas such as deforestation, sustainable forest management, biodiversity and illegal logging have faded into the background. Differing basic assumptions may in fact explain these conflicting attitudes toward the bio-economy concept. Pfau *et al.* (2014) distinguished scientists' views on the bio-economy, as either being assumed to inherently promote sustainability, or as being sustainable under certain, more or less, easily fulfilled conditions. And others describe the bio-economy as a cover concept for unsustainable forestry and forest industry practices. Pfau *et al.* noted that technology/science-oriented papers saw the bio-economy as being favorable for sustainability, whereas social science studies identified more complicating factors (Pfau *et al.* 2014).

By definition, the bio-economy is characterized as an economy that substitutes bio-based, renewable materials and services for fossil resources. In official documents, the bio-economy is, or is expected to be, driven by innovation of new superior products in a market economy. In this view the bio-economy is seen as having the potential to become a success story, but only under the proviso that its supply chains are genuinely sustainable and the final bio-products are able to provide customer value that can outcompete alternatives. Hence, to avoid overhasty and ungrounded conclusions, we do not imply that the bio-economy by definition is economically viable or sustainable – this must be specified, explained and finally confirmed by facts.

In the subsequent section we describe the potential for an increased contribution of the forest sector in a bio-economy. This is followed by an overview of the economic, policy and sustainability-related challenges for reaching this outcome. The final section synthesizes and provides (tentative) scenarios for a pathway to a forest sector that truly supports a bio-economy.

Current status of the bio-economy

To assess the forest sector's potential contribution to a bio-economy, the current trends for the forest sector in the world economy should first be gauged. What products and services are being released to the market today? What is the sector's share of the world economy and is that share growing or shrinking? The answers to these questions provide the departure point for an increasing significance of the forest sector in a bio-economy. And since the forest sector's support of a bio-economy also hinges on its sustainability record and competitiveness, these aspects also must be appraised.

Forests provide many goods and services, but the global economic reliance on forest products has decreased over time. Forests play a crucial role in providing renewable raw materials for products, energy and environmental services (UNEP 2011a, 151–152; EC 2012, 4). Global wood harvests are reaching 3.2 billion m^3 annually, corresponding to 0.7% of the growing stock. Of this, about half (1.5 billion m^3) is used for woodfuel and the other half for industrial wood, mainly sawnwood and pulpwood. However, important differences can be noticed between regions: 90% of the harvests in Africa are for energy whereas for North America the figure is only 7% (FAO 2010, xxi). Although essential for daily needs, this use of (affordable) wood for fuel also causes pollution with ensuing serious health consequences (Fullerton *et al.* 2008; IEA 2012, 14; Gordon *et al.* 2014). Between 600 and 800 million families, particularly women and children, are exposed to air pollution from plant-based fuels or charcoal, increasing respiratory infections, pneumonia, asthma and lung cancer, which may have killed 3.5–4 million people globally in 2010 (Gordon *et al.* 2014). Traditional forest industry products – lumber, pulp, paper, boards and energy – are mostly traded in mature markets and the global increase in wood use is still lower than global economic growth and more in pace with population growth. Statistics show a flat trend for sawnwood production (7% increase, 1994–2013) while paper increased by 48% over the same period, although global production of printing paper has been decreasing since the early years of the twenty-first century. The economic downturn and the financial shocks in 2008, have impacted the demand for sawnwood in the building sector in the important European and North American markets (PricewaterhouseCoopers 2013, 9; UNECE/FAO 2013, 11–12). For comparison, steel production in the last two decades (1994–2013) increased by 127% and oil production increased by 25% (World Steel Association 2015; FAO 2015; Energy Information Administration 2015). Global figures indicate that the main forest industries are mature – and there are other industries and sectors that are growing faster. Hence, economic data do not suggest an imminent booming forest-based bio-economy or fundamentally improved competitiveness of traditional forest products (through cost reduction or innovation). Improvements are, of course, constantly happening in the wood, pulp and paper industries but they are to a large degree gradual and in pace with innovation trends in other sectors. The job-creating capacity of the forest sector is gradually decreasing as larger, less labor intense production units are being built, and productivity is improving both in forest operations and processing. Employment in the forest sector is showing a decreasing trend from 16 million in 1990 to less than 13 million in 2006, representing only 0.4% of the global labor force. The sector's contribution to global GDP is about 1%, and this share appears to be decreasing although the production volume of the forest sector is stable or slightly increasing (FAO 2008, 41; FAO 2014, 20).

Although the forest sector plays a modest role in the world economy in official charts, it still generates plenty of informal income opportunities representing 41 million additional jobs, particularly in developing countries (UNEP 2011b, 8; FAO 2014, 18). Additional jobs are also generated in wood-based secondary workshops. The people living in forest areas in developing countries more frequently belong to the poorest and most marginalized groups, meaning that even if the absolute monetary value of forest products is low, they can constitute a crucial source of revenue and/or livelihood for many households, and a safety net in case of calamities (Sunderlin *et al.* 2008; Tesfaye 2011). The socioeconomic benefits from forests are imprecise but a rich pool of case studies show that forest non-timber income is of key importance for rural livelihoods in developing countries by providing food, honey and beeswax, and other products. In 2005, the value of non-wood forest products extracted from forests worldwide was estimated at US$18.5 billion (FAO 2010, 100). Other non-wood benefits from forests, both in developing and industrialized countries, include generic material, watershed regulation, climate regulation, recreation and cultural values (UNEP 2011b, 159).

One frequently omitted downside of small-scale forest use in developing countries concerns the cases where it may, due to necessity and poverty, erode the potential for future forest production. This fear may however be overstated as the main share of forest decline may be caused by land transformation (NASA Earth Observatory 2014). To put it more concretely, fuelwood collection may not cause as much deforestation as land use changes (Hosonuma *et al.* 2012), and poor rural households, although they are more forest dependent, use less forest resources than relatively more well-off households (Angelsen *et al.* 2014). In terms of sustainable forest management indicators the forest sector shows a very mixed picture (FAO 2010, 184–185). Biodiversity hotspots are being threatened in some regions while progress is noticed in terms of protected forests, the application of management plans and the amount of educated staff in forest organizations. Environmentally certified forests are also increasing, forming for the first time 10% of the global forest area (UNECE/FAO 2013, 17, 19). All this suggests that the world's forests may not yet be living up to their potential to contribute to a sustainable bio-economy – either in productive or sustainability terms (FAO 2010, 184–185). Deforestation is still progressing (FAO 2010, xxix), especially in poor countries where corruption is widespread. Ecosystem services are being threatened (Shvidenko *et al.* 2005, 607; TEEB 2010, 16). Illegal logging accounts for 15–30% of global timber production and over 50% in certain regions (the Amazon Basin, Central Africa and Southeast Asia) (Canby *et al.* 2013, 25–26). Despite efforts to thwart this through trade regulation there are still unsolved challenges in achieving considerably increased production of forest-based bio-products that are profitable, competitive and sustainable.

Referring to the three criteria for a forest-based bio-economy – increased use of forest resources, sustainability and economic viability – forests provide for many human needs and have important significance for a growing population and probably for the livelihoods of poor households. However, their relative importance is slowly diminishing and some products are quite stable and in some cases (e.g. newsprint) in decline. Forest products face intense competition in the marketplace from fossil-based products (for energy and plastics), concrete and steel (for construction), and information technology (in the case of newsprint). In addition, the global forest sector is still struggling with issues of sustainability and human health.

Potentials

Forest production

The contribution of forests to the bio-economy is, of course, determined by the capacity to produce forest raw materials that can be processed to compete with, and substitute for, non-renewable materials in a variety of applications. Forests cover about 30% of the world's land surface. This area is decreasing on a global scale and at an alarming pace in some countries, whereas other countries have managed to create a more stable situation, because of focused actions against deforestation and for reforestation, and economic and demographic transformations. The swift reduction of deforestation in China was mainly the outcome of fast urbanization, a "logging ban" after the major floods in 1998 and large-scale forest plantations. The production potential of forests to provide raw materials is furthermore restricted by the need for protected forests. Only 30% (1.2 billion hectares) of the global forest area is primarily designated for productive purposes, while a significant share (949 million hectares) has protective purposes. Large forest areas provide multiple benefits including timber production and protection (FAO 2010, 184–185). The increasing area of planted forests, 264 million hectares, or 6.6% of the global forest area, is gaining economic importance. These plantations are, however, a mixed blessing since they may replace natural forests with monocultures, and thereby may negatively impact biodiversity or local peoples (described in Chapter 9). However, plantations can also increase the production of feedstock (e.g. for new fuels and chemicals), and lower the pressure on natural forests. Forest area is not currently expanding, so increased timber use from forests, triggered by a radical transition to a bio-economy, may compete with the provision of other ecosystem services from forests. Agriculture will probably not be able to contribute much more to new bio-based, non-food products since agricultural land (37% of the world's land area) will be used to meet the increased demand for food (Haberl *et al.*, 2007; Kircher 2012b). An increased use of bio-based feedstock for energy and food that the bio-economy requires would therefore increase pressure on forest land.

The importance of forests as a non-cereal feedstock for bio-based products will grow. Wood is the most abundant non-food biomass input, globally accounting for 527 billion m^3 (131 m^3/hectare) making up 44% of the earth's biomass and it fixes a large share of the carbon that is sequestered each year (Kircher 2014). So, how can existing or decreasing forest areas with production restrictions provide the bio-resources for a bio-economy? Can this supply be modeled? Unfortunately, the future raw material supply from forests is difficult to predict based on historical data. Due to new technologies or fundamental changes in the global forest sector, recent conventional outlook studies have simply not proven to be accurate (Hurmekoski and Hetemäki 2013). Substantial effects can however be expected, for example due to a hypothetical drastically increased fuelwood demand (Buongiorno *et al.* 2010), or other market shifts. Buongiorno *et al.* (2012) analyzed one scenario based on Intergovernmental Panel on Climate Change (IPCC) scenarios assuming considerably increased demand for biomass energy, generating a 5.5 times higher wood energy demand. This would lead to increased roundwood prices and decreased use of roundwood for pulp and construction. World roundwood consumption would reach 11.2 billion m^3 in 2060 and wood prices would surge and exceed sustainable harvesting levels. On the other hand, the authors claim that with more stable fuelwood demand, consumption would only be 3.6 billion m^3 which may be compatible with sustainability criteria. Their study also indicates that biomass hypothetically could satisfy 14% of the world's primary energy needs. A reasonable estimate would be that world forest harvests will increasingly come from planted forests, whereas harvests from natural forests peaked in 1989 (Warman 2014). However, the key question concerning woody biomass use "is not the amount of resources, but rather their price" (Lauri *et al.* 2014).

Can a boosted global wood supply fit with sustainability requirements and cost restrictions? This capacity can, in theory, be enhanced through improved silvicultural methods, tree breeding, genetically modified trees (GMT) and fertilization. Changed rotation ages, fertilization and species selection increased forest production by 19% in mid-northern Sweden (Poudel *et al.* 2012). However, this may create conflict with environmental interests and public sentiment about genetic engineering, or with other land uses. A scramble for productive land may also call for various regulations and counter-measures including zoning, protection, and forest and agricultural intensification (Lambin and Meyfroidt 2011).

The industry

The traditional sectors of energy, sawnwood, and pulp and paper, will for some time constitute the backbone of the forest sector's role in a bio-economy. Increased productivity and product diversification – and also mundane daily improvements – are critical for a successful contribution by the forest sector to a bio-economy. There are many technologies available today that can significantly and positively impact forestry operations. For example, digital terrain models and geographical positioning systems for harvest planning for reduced ground damage and improved productivity, laser scanning for timber calculations, and satellite image analysis for inventories. These technologies offer opportunities for both more productive and more agile forest-based value chains, costs savings and enhanced product quality. At the same time, adoption often means a continued exodus of forest-based work opportunities away from forested, rural areas.

The forest industry has become more efficient, for example through optimized processes (e.g. value optimization in dry sorting lines, better monitoring of processes, more efficient energy use and improved output ratios). Still, compared to other sectors, innovations and improvements are slow, process-oriented and incremental. Future improvements downstream in the wood market involve new industrial building systems based, for example, on cross-laminated timber with the advantage of allowing tall buildings, industrial methods and better documentation of environmental advantages (carbon storage). Glulam represents low costs, fast building times and desired strength properties, and is increasingly employed in large halls, sports facilities and warehouses (UNECE/FAO 2013, 106). Industrial building approaches have been successfully applied to streamline the whole building process while more activities are localized indoors in controlled conditions instead of in a wet environment (Sardén 2005, 21; Brege *et al.* 2013). Improvements are also affecting retailing and logistics (e.g. through increased e-trade of forest products).

Due to an expected continued fall in demand for paper for printing purposes, the pulp and paper industry is forced to continue a constant improvement of processes and move outside its comfort zone into new products and applications. However, growth in different paper segments is seen in fast-growing regions and new e-trade and retailing methods are still economically important for the packaging segment. Gradual improvements in the pulp and paper industry include automation, process control, more sophisticated use of sensors and more efficient and energy-saving production. The industry strives to improve energy efficiency and gradually improve different paper products in terms of strength and printability and development of "intelligent paper" (Heikenfeld *et al.* 2011).

Energy

One of the largest, low-cost uses of the bio-resource has been and will continue to be energy. This use may continue to grow from 50 EJ today to 160 EJ in 2050, making bioenergy one of the most important renewable energy sources. Although bioenergy is most appropriately used for heating, its use for the generation of electricity could prevent 1.3 gigatonnes of CO2 equivalents being emitted into the atmosphere by 2050 (IEA 2012, 5). The use of biomass for electricity can furthermore be attained through co-firing with coal in combined heat and power plants. Gasification and subsequent electricity production is yet another technology under development.

Much of the traditional use of wood and residues for cooking and heating is of low efficiency, has negative health effects and is not especially sustainable. Hence, new efforts should address these issues that in the end could turn the large sector of domestic bioenergy into a more attractive energy form. Other improvements are in the processing and upgrading of bioenergy carriers, and include drying, pelletization and briquetting, which transform the material to a more energy-dense and homogeneous fuel resource. Further treatments are torrefaction into tar and even more energy-rich pellets, and pyrolysis up to 400–600°C can generate fuels with an even higher energy density (IEA 2012, 13).

In some European countries, wood energy is a key component in energy policies and goals, such as the EU 2020 goal stating that renewable energy sources should account for 20% of gross energy consumption by 2020. Several countries have, through policies and economic incentives, increased the use of wood residues for energy and heat. The next stage involves increasing use of liquid biofuels where the target is set at 10% for 2020.

First-generation fuels (i.e. biodiesel, bio-esters, bio-ethanol and biogas) form 3% of total road transport fuel globally (IEA 2011, 11) but suffer from drawbacks because of competition for farmland and sustainability and economic issues – perhaps with the exception of sugarcane-based ethanol (Sims *et al.* 2010). "Second-generation" biofuels are based on a much more abundant biomass (i.e. lignocellulosic feedstock), and can potentially offer more sustainable and climate-neutral processes. Scale and efficiency improvements will hopefully reduce the production costs of these biofuels over time (IEA 2011, 31). For these biofuels to become truly competitive capital requirements must be reduced as well as costs controlled through scaling up and learning. Then, biofuels can, together with substantial efficiency improvements, contribute significantly to a constant level of CO2 emissions from the transport sector despite a large increase of the vehicle fleets in developing countries. According to the IEA (2011, 43) important increases in the consumption of biofuels by 2050 will take place in developing countries. Large energy companies are investing heavily in biofuel programs (Sims *et al.* 2010) and globally, biofuels could provide 27% of total transport fuels by 2050, avoiding another 2.1 gigatonnes of CO2 emissions per year (IEA 2011, 21).

Chemicals and materials

Annual oil production is approximately 3.9 billion tons, and 92% of this is used for energy and 8% for chemicals. The share of biochemicals was 3–4% in 2010, but it is expected to grow to 7–17% in 2025 and further after that (Kircher 2014). In connection with the ambition to create an innovative forest sector in a bio-economy, opportunities therefore exist to diversify and develop new chemical products. Bio-based chemicals include, besides regular by-products in the pulp and paper process and biofuels, different product classes from bulk products, specialty and fine products, and highly expensive pharmaceuticals (Kircher 2014). Bozell and Petersen (2010) and Posada *et al.* (2013) highlight chemicals from carbohydrates with commercial potential for the following: energy, fuel, solvents, antiseptics, basic compounds for specialty chemicals, pharmaceuticals, industrial chemicals with many uses, softeners, food additives, bases for producing bioplastics, plasticizers and sweeteners. However, Bozell and Petersen (2010) warn that the set of promising compounds will constantly change and the most promising money generators depend on how they can provide market share and business prospects for investors.

Wood plastic composites (WPCs) present good growth prospects and are the basis for several construction systems, furniture and decking. The bio-component of WPCs contributes to a bio-economy although some varieties also include fossil-based components (UNECE/FAO 2013, 27). Use of bio-based polymers, for natural-fiber-reinforced composites (NFC) is projected to grow many times over, until the late 2010s, promoted by reduced oil supply and environmental concerns (UNECE/FAO 2013, 23; Philip 2013). The reinforcement of polymers by natural fibers has been used since the first half of the twentieth century and "green bio-composites" are materials where both the matrix and the fibers are bio-based. These biopolymers constitute a range of "building blocks", which can be produced with highly desired properties – with or without the ability to be biodegradable. Important bio-based materials also include PLA (polylactic acid, which can replace polyethylene terephthalate), starch plastics, bio-based polyethylene and PHAs (polyhydroxy alkanoates). These bio-materials may enter new environmentally sensitive markets such as cars and boats (Zini and Scandola 2011; UNECE/FAO 2013, 32; Mizrachi *et al.* 2012; Fowler *et al.* 2006). The development of new materials is propelled by several driving forces, such as environmental concerns, recycling directives and the search for high performance in terms of density, toughness and biodegradability. Wood-based textile thread is spun from viscose solution made from dissolving-pulp and can be used to produce viscose textiles for clothing, but also specialty textiles. For environmental, water conservation and climate reasons, as well as their attractiveness, cellulosic fibers are seeing increased interest from the fashion industry (Shen *et al.* 2010). Dissolving-pulp can also be used in such different applications as sausage casing, tire cord and as a component in fillers in pharmaceutical tablets. Durapulp is a combination of cellulose and bio-polymer that can be pressed into two- or three-dimensional shapes. The material resembles paper, but has other characteristics such as high folding strength, high tear strength, high bending stiffness, high air permeability, high dimensional stability and low water absorption and can be adapted to suit a number of applications (Roos *et al.* 2014).

However, improving forest-based materials' performance and functionality is not sufficient for economic and market success. To make economic logic, they must achieve high volumes, and processes must be improved through a learning process "learning by running" – and not least, customer preferences must be understood (Kircher 2012b). This creates new challenges to the traditional forest companies to forge collaborations with downstream industries, such as the car industry or manufacturing that embrace different business models and logics.

Biorefineries

Newer, innovative bio-based products, are generally outcomes of biorefinery production using pyrolysis, hydrolysis, biomass gasification, or other thermal, chemical or biological processes (UNECE/FAO 2013, 28; Fitzpatrick *et al.* 2010; Naik *et al.* 2010). Processing steps include, biofuel choice, pretreatment, production of intermediary products and final products. The effectiveness of enzymatic processes represents one hurdle to more intensified production of chemicals and fuels (Menon and Rao 2012). Biorefinery processes provide different products such as construction materials, additives, paints and also fuels and energy (UNECE/FAO 2013, 28; Sandén and Pettersson 2013). The products range from high-value-added specialty chemicals and materials to high-volume, low-value-added energy carriers. Biorefineries ideally combine production with innovation and development (Fitzpatrick *et al.* 2010; Centi *et al.* 2011).

Different biorefinery platforms require specific biomass to produce a range of products (IEA-Bioenergy 2012, 6–8). The challenge for biorefineries is to overcome the higher feedstock costs compared to average oil extraction costs and to design appropriate supply systems (Kircher 2014; Mizrachi *et al.* 2012). In the fossil-based economy, crude oil, which is a quite homogeneous raw material, is processed in very large refineries close to large markets. Due to the dispersed nature of biomass, biorefineries will need to be located closer to the feedstock and likely be of a smaller scale than petrochemical refineries (Kircher 2012b). Important adaptations in infrastructure are needed to create cost efficient and coordinated supply chains (Kircher 2012a). The successful development of biorefineries involves good market assessments, a wise adaptation to existing local facilities and infrastructure, quick learning, and a proper evaluation of sustainability effects (Wellisch *et al.* 2010).

Challenges

The economic challenges

Economic incentives are influencing the forest sector's contribution to a sustainable and economically sound bio-economy. Significant investments in forest biomass capacity will continue until 2050 although cost differences between fossil energy and biomass energy will remain a challenge (IEA 2012, 5). The largest bioenergy cost is represented by feedstock for conventional fuels (45–70%) although for advanced fuels, capital costs will become increasingly important. It is therefore crucial to make use of by-products and extract the full economic value of co-production processes in biorefineries.

Is a sustainable bio-economy realistic and feasible on a global scale? An increasing global population – mainly concentrated in cities in developing countries (United Nations Department of Economic and Social Affairs 2014) – that furthermore becomes more integrated into the world economy, will inevitably increase the demand for bio-based products. More sustainable and efficient forest products value chains can help meet this population's increasing needs, while simultaneously moving away from fossil fuel use. However, the new opportunities are not yet manifesting themselves in increasing demand for forest products or increasing biomass capacity. The forest industry is instead being squeezed by low economic growth in industrialized countries, high timber prices, slow technology transformations, overcapacity and low profitability (PricewaterhouseCoopers 2013, 5).

Customer behavior could potentially also support an emerging bio-economy. This in turn warrants high-quality bio-based products, and a knowledge and awareness among producers about customer needs. The promotion of environmental customer choices highlights a dire need for environmental performance metrics and certification systems. According to the Ecolabel Index (2015) 459 ecolabels currently co-exist, confusing buyers and rendering decision-making and handling of certificates complicated and costly (Räty *et al.* 2012, 42).

The challenges for biochemicals and biofuels are associated with production and handling costs, slightly more complicated processing steps, transport and logistics. Production must also become streamlined and strategically appropriate for the intended end products and supply system. A wise, yet versatile, combination of different output products must be conceived, which will require new collaborations and partnerships between different industry sectors.

Policy challenges

Policies for a sustainable bio-economy must support the substitution of bio-based products for non-renewable products. Energy policies, certificates, taxes and fees can reduce the competitiveness of fossil energy and support the emergence and growth of bio-based energy where environmental externalities are internalized in the production costs. Forest policies can help to promote sustainable forest management and innovation policies can promote commercialization of new ideas. Policies must also serve to regulate conflicts in society, for example between production and conservation concerns. This involves handling conflict between an increased use of bio-based resources on the one hand, and further developing new products and markets on the other. New demand for wood caused by new bio-based products might result in tensions within the forest sector, for example between the traditional forest industry and bioenergy buyers. Disputes can also occur between forestry production and conservation interests focusing on ecological goods and services. Forest management today is mainly regulated by national policies and international commitments (Puddister *et al.* 2011). As the increasing world population will engender the flow of more forest-based and agricultural products, alongside more ecosystem services, appropriate reconciling policies that take into consideration how land use interacts with other sectors (e.g. for employment) and the global economy (concerning trade flows) (Lambin and Meyfroidt 2011) are needed. It is crucial that the bio-economy is not developed on the backs of the poorest farmers in developing countries.

Many countries have implemented bio-economy strategies and policies aimed at decoupling economic growth from environmental degradation. However, Staffas *et al.* (2013) observe a striking variation in how policies are envisioned and a lack of focus in how the concept is defined and which policies would lead to the end result. The bio-economy policy process will certainly highlight several of the conflicts we have today around different perceptions of wood, the urban–rural divide, market forces, etc., expected by Richardson (2012).

The ambition to increase the use of bio-resources targets a range of uses: materials, biochemical production, biofuels and biomass for heat and power (Kircher 2012a). Hence, national policies must support investments, entrepreneurs, the development of a skilled workforce and policies for innovation in these areas. Promoting an economically competitive bio-economy requires that these actions be well targeted indeed to enable the establishment of appropriate infrastructure, efficient value chains and the right kind of private–public partnerships (Kircher 2012a). Policies for boosting the bio-economy may involve implementation of climate policies (e.g. carbon markets) and non-discriminatory regulations for materials that open new applications for wood in buildings (UNECE/FAO 2013, 29–30). The establishment of bio-refineries is best supported by subsidies for investments and support for innovation and development (Hämäläinen *et al.* 2011). Still, these investments will continue to be hampered by the sheer size of the capital investments required.

Summarizing key documents, a number of policies emerge as key drivers to achieving a forest-based economy, globally: support of research and development, creating startups and an innovative business environment, providing the proper environment for a bio-based industry meaning that it is supplied with stable institutions, talent and know-how, and sound economic incentives. Required policy shifts include removal of fossil fuel subsidies, support of new bio-technologies in their early stages, creation of a stable policy framework, introduction of bio-products sustainability indicators that are linked to the products' contribution to life cycle CO2 emissions and sustainability impacts, promotion of trade in bio-products and help for the industry to solve complex supply chain and infrastructure issues.

The sustainability challenges

The bio-economy concept generally implies sustainable resource use. However, referring to Pfau *et al.* (2014), sustainability is not an intrinsic property of an economy based on bio-resources, without qualification. Instruments are available to measure and monitor sustainability criteria. Forest certification and due diligence measures aim to combine economic use with sustainability (UNECE/FAO 2013, 18, 105). Further options include building systems such as LEED and BREAM that offer green features (UNECE/FAO 2013, 22). Other criteria and standards for sustainability are: Global

Bioenergy Partnerships, the Roundtable on Sustainable Biofuels, ISO, and the International Sustainability and Carbon Certification System. These systems and indicators have, however, not yet reached their potential to become transparent mechanisms that help industrial and end-customers to easily make the best choices.

There is still a lack of awareness in policy documents about the balance of tradeoffs between sustainability criteria pertaining to forests and economic aspirations for a bio-economy. Despite advances, there is also a lack of criteria and indicators on how success should be measured. It takes more effort to keep the forest sector at the forefront of the transition to a bio-economy. If a bio-economy requires high-yielding forest production over vast areas, the sustainability challenge for a bio-economy is not as uncontroversial as it may seem.

Synthesis of potential and challenges

The potential sustainability and competitive challenges to support a bio-economy, for different forest sector products and for land use, are presented in Table 10.1.

The table summarizes issues and the potential for specific forest-based sectors to support a sustainable bio-economy. It identifies certain traditional sectors, mainly paper and wood products, where the potential additional innovation may only be gradual as will be the increased contribution to the bio-economy. These sectors do, on the other hand, represent the backbone of the forest sector's present industrial contribution to a bio-economy, and a base for further advances. The development of bioenergy use may have greater potential in some regions if sustainability issues are addressed. Liquid biofuels offer potential bio-economy contributions as they may reduce fossil fuel dependence. This development does, however, require further development of processing techniques. Tight competition and conflicts are imminent if land use needs to become more intensified. Zoning and a careful balance of the forest's production of feedstock, ecological services and income for local dwellers, must be attained.

Conclusion: The forests' role in a bio-economy

If the vision of a forest-based bio-economy is to become a reality and not a mirage that dissipates when all facts become clear, we believe it should address a few key conditions for long-term success. Otherwise there is a danger the concept is headed for a "hard landing", which may prevent good ideas from being realized at all. These comments should not be taken as gloomy negativism, but are rather founded in a belief that the forest sector can contribute to a vibrant and sustainable bio-economy. Our key messages are concentrated in a few statements below:

Table 10.1 Bio-economy potential for forest-based product groups

Product	*Economic potential and competitiveness*	*Sustainability issues*	*Action needed for economic and sustainability performance*
Paper	Mainly mature product groups of high economic importance (still) for the forest sector. Printing paper is on the decrease and packaging is growing. Incremental product and process improvements	High energy input required. Environmental and resource efficiency is improving through better processing methods. Unsustainable logging methods remain in some regions but are also being prohibited	Further improve quality and resource efficiency in production. Continue to phase out unsustainable/illegal supply chains
Sawnwood	Mature product group where new processes, product offerings and building methods can be gradually developed	Wood generally has environmental and climate advantages compared to steel and concrete. Unsustainable supply chains remain (see above)	Processing and building methods and products can be developed. Wood construction has environmental advantages. Continue to phase out unsustainable/illegal supply chains
Bioenergy	Scale effects can make wood-based bioenergy efficient but not cost-competitive with, e.g., coal, except in certain segments, if externalities are not accounted for. Wood energy is crucial for households and employment in developing countries, but it involves health problems	Using sustainably grown wood biomass instead of fossil energy creates direct sustainability gains. Smoke pollution in developing countries is a health issue	Competitive and sustainable use of biomass for heating is feasible with the right type of economic incentives. Further development of bioenergy offerings and supply chains. Sustainable bioenergy use in developing countries can be promoted more (improved stoves)

Product	*Economic potential and competitiveness*	*Sustainability issues*	*Action needed for economic and sustainability performance*
Liquid biofuels	First-generation biofuels could be competitive. Second-generation biofuels are being developed and may soon be ready for full-scale production	First-generation biofuels compete with food crops. Second-generation biofuels based on lignocellulose will provide better sustainability results	Second-generation biofuels may present technical- and sustainability-related advantages and replace fossil transportation fuels. Subsidies to fossil energy should stop
Chemicals	New products are developed in biorefineries. Growth potential which can bring economic revenues but less employment	Bio-based chemicals can replace compounds based on petroleum. Although the quantities are not large enough to have a great climate effect	New products can be produced but their total impact on the bio-economy is modest. Biochemicals involve promising innovation areas
Bioplastics/composites	New products are developed in biorefineries. Although most are in the development stage	Bio-based chemicals can replace materials based on fossil oil. The quantities are not large	New high value products can be developed. Their total impact on the bio-economy is modest. Important field for R&D
Land use	A potential transition to a bio-economy will increase pressure on both agriculture and forest land. Competition with other land uses may intensify. Primary production may need to be intensified	Forest biomass does not compete with food production. Intensified forest use, plantations, and genetically modified trees may create conflicts with conservation interests. Extensive use of forest biomass may equate to higher carbon sequestration in forests. Wide range of policy measures needed to safeguard sustainability and defend livelihoods of poor rural people	Forest management's contribution to a bio-economy is possible but may involve tricky tradeoffs and conflicts. Efficient, yet sustainable, production systems must be further developed

- *Win the customer.* Bio-based alternatives (fuels, materials, chemicals, etc.) may offer high performance but the key proof is if they succeed in providing superior benefits to the customer, who may be an industrial or private buyer. Their choices are in turn based on tradition, experience, risk considerations, policies – and primarily price. The key message to the forest sector is that their products, after all these aspects have been taken into consideration, must end up on top. The forest sector must understand the market and the customers' needs. This statement is a truism, we agree, but it is nevertheless valid.
- *Understand the overall production logics.* Providing these benefits, for a reasonable cost, requires efforts on a wide range of aspects to get the production network right. An integrated approach is surely needed that synchronizes technological improvements, uses a sensitive market radar including trans-sectorial contacts, creates global supply networks, and utilizes a supportive infrastructure (Kircher 2012b). The forest-based bio-economy is furthermore an open system interacting with other sectors, land uses and other receivers of forest benefits. Its destiny is tied to other sectors and the global economy. Hence, the key actors in the forest sector must understand this and learn to communicate with and develop ideas together with other industries and stakeholders.
- *Don't take the blessing of a forest bio-economy for granted.* The key importance of the forest sector for a bio-economy must be proven. High-flying visions may be needed and assertiveness too. And yes, forest-based products have a unique position for underpinning a sustainable bio-based economy. Still, there are things to prove in terms of sustainability performance and competitiveness. Industry-wide complacency is a serious fault and other substitute sectors are fiercely improving both product quality and sustainability performance. The environmental performance of forest activities and products must be improved further and transparent indicators should be developed and communicated to the customer.
- *Spot the current and looming conflicts.* There are several hidden contradictions that may prove more critical if the bio-economy becomes a reality. Its development will warrant a demand for highly productive and cost-efficient production systems for biomass and raw materials that may impinge on biodiversity and other values. This calls for more efficient land use for both conservation and productive purposes. How can this be achieved? Experts highlight the potential conflicts embedded in the introduction of a bio-economy, advising adaptive management, monitoring, and constant review of management regimes based on science (Puddister *et al.* 2011). Zoning and improved labor markets will also influence land use in developing countries (Lambin and Meyfroidt 2011).
- *It's a local and global process.* The North–South difference merits specific attention, although this division of the world is fast becoming obsolete. If the bio-economy concept is to be taken seriously, sections of the forest-based economy that still are harmful for health or are

unsustainable must be addressed while acknowledging the significance of forests for poor communities. The development of a bio-economy must conform to sustainable development. To be real, the bio-economy must prove that it is sustainable and competitive and that it can provide superior fuels, materials, products and services. It needs coordination and systemic changes and it requires infrastructure and long-range planning of facilities.

- *Attract talent.* To reach this stage, the sector needs to build more knowledge and development skills. Good research and innovation is needed based on trans-disciplinary approaches that join the biological, technical and social sciences. However, maybe more importantly, the bio-economy depends on talented people to develop new products, and organize the flows and investments. A true understanding of innovation is needed to build a bio-economy based on competitive advantage and high sustainability standards.

References

Angelsen, A., Jagger, P., Babigumira, R., Belcher, B. (2014) 'Environmental income and rural livelihoods: a global comparative analysis', *World Development*, 64, pp. S12–S28.

Bozell, J.J. and Petersen, G.R. (2010) 'Technology development for the production of biobased products from biorefinery carbohydrates – the US Department of Energy's "Top 10" revisited', *Green Chemistry*, 12(4), pp. 539–554.

Brege, S., Stehn, L. and Nord, T. (2013) 'Integrated design and production of multi-storey timber frame houses: production effects caused by customer-oriented design', *Int. J. Production Economics*, 77, pp. 259–269.

Buongiorno, J., Raunikar, R. and Zhua, S. (2010) 'Consequences of increasing bio-energy demand on wood and forests: an application of the global forest products model', *Journal of Forest Economics*, 17, pp. 214–229.

Buongiorno, J., Zhu, S., Raunikar, R. and Prestemon, J.P. (2012) 'Outlook to 2060 for world forests and forest industries: a technical document supporting the Forest Service 2010 RPA assessment', U.S. Department of Agriculture Forest Service, Southern Research Station. Technical Report SRS-151, Asheville, NC.

Canby, K. and Oliver, R. (2013) *European Trade Flows and Risk.* Forest Trends, Washington DC and Forest Industries Intelligence Limited, UK.

Centi, G., Lanzafame, P. and Perathoner, S. (2011) 'Analysis of the alternative routes in the catalytic transformation of lignocellulosic materials', *Catalysis Today*, 167, pp. 14–30.

Enriquez-Cabot, J. (1998) 'Genomics and the world's economy', *Science Magazine*, 281, pp. 925–926.

Duchesne, L.C. and Wetzel, S. (2003) 'The bioeconomy and the forestry sector: changing markets and new opportunities', *The Forestry Chronicle*, 79(5), pp. 860–864.

EC (2012) *Innovating for Sustainable Growth: A Bioeconomy for Europe.* European Commission, Brussels.

Ecolabel Index (2015) 'Global directory of ecolabels'. www.ecolabelindex.com, accessed 5 Aug. 2015.

Energy Information Administration (2015) 'Production of crude oil including lease condensate (thousand barrels per day)'. www.eia.gov/cfapps/ipdbproject/iedindex3.cfm?tid=5&pid=57&aid=1&cid=ww,&syid=1993&eyid=2013&unit=TBPD, accessed 5 Aug. 2015.

FAO (2008) 'Contribution of the forestry sector to national economies, 1990–2006'. Working paper: FSFM/ACC/08. Food and Agriculture Organization of the United Nations, Rome.

FAO (2010) 'Global forest assessment 2010'. Forestry paper: 163. Food and Agriculture Organization of the United Nations, Rome.

FAO (2014) 'State of the world's forests. Enhancing the socioeconomic benefits from forests', Food and Agriculture Organization of the United Nations, Rome.

FAO (2015) 'FAOSTAT – Forestry'. http://faostat3.fao.org/download/F/FO/E, accessed 5 Aug. 2015.

Fitzpatrick, M., Champagne, P., Cunningham, M.F. and Whitney, R.A. (2010) 'A biorefinery processing perspective: treatment of lignocellulosic materials for the production of value-added products', *Bioresource Technology*, 101, pp. 8915–8922.

Fowler, P.A., Hughes, J.M. and Elias, R.M. (2006) 'Biocomposites: technology, environmental credentials and market forces', *Journal of the Science of Food and Agriculture*, 86(12), pp. 1781–1789.

Fullerton, D.F., Bruce, N. and Gordon, S.B. (2008) 'Indoor air pollution from biomass fuel smoke is a major health concern in the developing world', *Transactions of the Royal Society of Tropical Medicine and Hygiene*, 102, pp. 843–851.

Gordon, S.B., Bruce, N.G., Grigg, J., Hibberd, P.L., Kurmi, O.P., Lam, K.H., Mortimer, K., Asante, K.P., Balakrishnan, K., Balmes, J., Bar-Zeev, N., Bates, M.N., Breysse, P.N., Buist, S., Chen, Z., Havens, D., Jack, D., Jindal, S., Kan, H., Mehta, S., Moschovis, P., Naeher, L., Patel, A., Perez-Padilla, R., Pope, P., Rylance, J., Semple, S. and Martin, W.J. (2014) 'Respiratory risks from household air pollution in low and middle income countries', *The Lancet Respiratory Medicine*, 2, 823–860.

Haberl, H., Erb, K.H., Krausmann, F., Gaube, V., Bondeau, A., Plutzar, C., Gingrich, S., Lucht, W. and Fischer-Kowalski, M. (2007) 'Quantifying and mapping the human appropriation of net primary production in earth's terrestrial ecosystems', *Proceedings of the National Academy of Sciences*, 104(31), pp. 12942–12945.

Hall, R., Smolkers, R., Ernsting, A., Lovera, S. and Alvarez, I. (2012) Bio-economy versus Biodiversity. Global Forest Coalition. http://globalforestcoalition.org/wp-content/uploads/2012/04/Bioecono-vs-biodiv-report-with-frontage-FINAL.pdf, accessed 31 Oct. 2014.

Hämäläinen, S., Näyhä, A. and Pesonen, H-L. (2011) 'Forest biorefineries – a business opportunity for the Finnish forest cluster', *Journal of Cleaner Production*, 19, pp. 1884–1891.

Hardy, R.W.F. (2002) 'The bio-based economy', in *Trends in New Crops and New Uses*. Janick, J. and Whipkey, A. (eds). ASHS Press, Alexandria, VA.

Heikenfeld, J., Drzaic, P., Yeo, J-S. and Koch, T. (2011) 'A critical review of the present and future prospects for electronic paper', *Journal of the SID*, 19(2), pp. 129–156.

Hosonuma, N., Herold, M., De Sy, V., De Fries, R.S., Brockhaus, M., Verchot, L., Angelsen, A. and Romijn, E. (2012) 'An assessment of deforestation and forest degradation drivers in developing countries', *Environmental Research Letters*, 7(4), pp. 1–12.

Hurmekoski, E. and Hetemäki, L. (2013) 'Studying the future of the forest sector: review and implications for long-term outlook studies', *Forest Policy and Economics*, 34, pp. 17–29.

IEA (2011) *Technology Roadmap – Biofuels for Transport*. International Energy Agency, Paris, France.

IEA (2012) *Technology Roadmap – Bioenergy for Heat and Power*. International Energy Agency, Paris, France.

IEA-Bioenergy (2012) *Bio-based Chemicals – Value Added Products from Biorefineries*. International Energy Agency Task 42 Biorefinery, Paris, France.

Kircher, M. (2012a) 'The transition to a bio-economy: emerging from the oil age', *Biofuels Bioproducts & Biorefining*, 6(4), pp. 369–375.

Kircher, M. (2012b) 'The transition to a bio-economy: national perspectives', *Biofuels Bioproducts & Biorefining*, 6(3), pp. 240–245.

Kircher, M. (2014) 'The emerging bioeconomy: industrial drivers, global impact, and international strategies', *Industrial Biotechnology*, 10, pp. 11–18.

Kleinschmit, D., Hauger Lindstad, B., Jellesmark Thorsen, B., Toppinen, A., Roos, A. and Baardsen, S. (2014) 'Shades of green: a social scientific view on bioeconomy in the forest sector', *Scandinavian Journal of Forest Sciences*, 29(4), pp. 402–410.

Lambin, E.F. and Meyfroidt, P. (2011) 'Global land use change, economic globalization, and the looming land scarcity', *Proceedings of the National Academy of Sciences*, 108(9), pp. 3465–3472.

Lauri, P., Havlík, P., Kindermann, G., Forsell, N., Böttcher, H. and Obersteiner, M. (2014) 'Woody biomass energy potential in 2050', *Energy Policy*, 66, pp. 19–31.

Menon, V. and Rao, M. (2012) 'Trends in bioconversion of lignocellulose: biofuels, platform chemicals and biorefinery concept', *Progress in Energy and Combustion Science*, 37, pp. 522–550.

Mizrachi, E., Mansfield, S.D. and Myburg, A.A. (2012) 'Cellulose factories: advancing bioenergy production from forest trees', *New Phytologist*, 194(1), pp. 54–62.

Naik, S.N., Goud, V.V., Rout, P.K. and Dalai, A.K. (2010) 'Production of first and second generation biofuels: a comprehensive review', *Renewable and Sustainable Energy Reviews*, 14, pp. 578–597.

NASA Earth Observatory (2014) *Causes of Deforestation: Direct Causes*. http://earthobservatory.nasa.gov/Features/Deforestation/deforestation_update3.php 21 October 2014.

OECD (2011) *Towards Green Growth*. Organisation for Economic Co-operation and Development, Paris.

Philip, J. (2013) 'OECD policies for bioplastics in the context of a bioeconomy, 2013'. *Industrial Biotechnology*, February, pp. 19–21.

Pfau, S.F., Hagens, J.E., Dankbaar, B. and Smits, A.J.M. (2014) 'Visions of sustainability in bioeconomy research', *Sustainability*, 6, pp. 1222–1249.

Posada, J.A., Patel, A.D., Roes, A., Blok, K., Faaij, A.P.C. and Patel, M.K. (2013) 'Potential of bioethanol as a chemical building block for biorefineries: preliminary sustainability assessment of 12 bioethanol-based products', *Bioresource Technology*, 135, pp. 490–499.

Poudel, B.C., Sathre, R., Bergh, J., Gustavsson, L., Lundström, A. and Hyvönen, R. (2012) 'Potential effects of intensive forestry on biomass production and total carbon balance in north-central Sweden', *Environmental Science and Policy*, 15, pp. 106–124.

PricewaterhouseCoopers (2013) *Global Forest, Paper & Packaging Industry Survey 2013 edition – survey of 2012 results*. Ontario, Canada.

Puddister, D., Dominy, S.W.J., Baker, J.A., Morris, D.M., Maure, J., Rice, J.A., Jones, T.A., Majumdar, I., Hazlett, P.W., Titus, B.D., Fleming, R.L. and Wetzel, S. (2011) 'Opportunities and challenges for Ontario's forest bioeconomy', *Forestry Chronicle*, 87(4), pp. 468–477.

Pülzl, H., Kleinschmit, D. and Arts, B. (2014) 'Bio-economy – an emerging meta-discourse affecting forest discourses?' *Scandinavian Journal of Forest Research*, 9(4), pp. 386–395.

Räty, T., Lindqvist, D., Nuutinen, T., Nyrud, A.Q., Perttula, S., Riala, M., Roos, A., Tellnes, L.G.F., Toppinen, A. and Wang, L. (2012) 'Communicating the Environmental Performance of Wood Products'. Working Papers of the Finnish Forest Institute 230, Vantaa, Finland.

Richardson, B. (2012) 'From a fossil-fuel to a biobased economy: the politics of industrial biotechnology', *Environment and Planning C – Government and Policy*, 30(2), pp. 282–296.

Roos, A., Lindström, M., Heuts, L., Hylander, N., Lind, E. and Nielsen, C. (2014) 'Innovation diffusion of new wood-based materials – reducing the "time to market"', *Scandinavian Journal of Forest Research*, 29(4), pp. 394–401.

Sandén, B. and Pettersson, K. (eds) (2013) *Systems Perspectives on Biorefineries 2013*. Chalmers University of Technology, Göteborg, Sweden. e-published at: www.chalmers.se/en/areas-of-advance/energy/cei/Pages/Systems-Perspectives.aspx.

Sardén, Y. (2005) 'Complexity and learning in timber frame housing. The case of a solid wood pilot project', PhD thesis, Luleå, Luleå University of Technology.

Shen, L., Worrell, E. and Patel, M. (2010) 'Environmental impact assessment of man-made cellulose fibres', *Resources Conservation and Recycling*, 55(2), pp. 260–274.

Shvidenko, A., Barber, C.V. and Persson, R. (2005) 'Forest Systems'. Ch. 21 in *Millennium Ecosystem Assessment Series, 'Ecosystems and Human Well-being: Current State and Trends'*, Volume 1 (eds Rashid Hassan, R. Scholes and N. Ash). Island Press, Washington DC.

Sims, R.E.H., Mabee, W., Saddler, J.N. and Taylor, M. (2010) 'An overview of second generation biofuel technologies', *Bioresource Technology*, 101, pp. 1570–1580.

Staffas, L., Gustavsson, M. and McCormick, K. (2013) 'Strategies and policies for the bioeconomy and bio-based economy: an analysis of official national approaches', *Sustainability*, 5, pp. 2751–2769.

Sunderlin, W.D., Dewi, S., Puntodewo, A., Muller, D., Angelsen, A. and Epprecht, M. (2008) 'Why forests are important for global poverty alleviation: a spatial explanation', *Ecology and Society*, 13(2), Art. 24.

Swedish Forest Industries Federation (2012) The forest industry – the driver for a sustainable bioeconomy – Sustainability Report 2012. Stockholm.

TEEB (2010) *The Economics of Ecosystems and Biodiversity: Mainstreaming the Economics of Nature: A synthesis of the approach, conclusions and recommendations of TEEB*, The Economics of Ecosystems and Biodiversity (TEEB), Progress Press, Malta.

Tesfaye, Y. (2011) 'Participatory forest management for sustainable livelihoods in the Bale Mountains, Southern Ethiopia'. PhD thesis, Acta Universitatis agriculturae Sueciae, Swedish University of Agricultural Sciences, Uppsala, Sweden.

UNECE/FAO (2011) The forest sector in the green economy, *Geneva timber and forest discussion paper 54*. United Nations Economic Commission for Europe/Food and Agriculture Organization of the United Nations, Geneva.

UNECE/FAO (2013) *Forest Products Annual Market Review*, United Nations Economic Commission for Europe/Food and Agriculture Organization of the United Nations, Geneva.

UNEP (2011a) *Towards a Green Economy: Pathways to Sustainable Development and Poverty Eradication*, United Nations Environment Programme.

UNEP (2011b) *Forests – Investing in Natural Capital*, United Nations Environment Programme.

United Nations (2012) The Future We Want, United Nations Conference on Sustainable Development, Rio de Janeiro, Brazil.

United Nations Department of Economic and Social Affairs (2014) *World Urbanization Prospects, 2014 revision*, United Nations, New York.

Warman, R.D. (2014) 'Global wood production from natural forests has peaked', *Biodivers Conserv*, 23, pp. 1063–1078.

Wellisch, M., Jungmeier, G., Karbowski, A., Patel, M.K. and Rogulska, M. (2010) 'Biorefinery systems – potential contributors to sustainable innovation', *Biofuels Bioproducts & Biorefining*, 4(3), pp. 275–286.

White House (2012) *National Bioeconomy Blueprint*, Washington DC.

World Steel Association (2015) 'Annual steel production 1980–2013'. www.worldsteel.org/statistics/statistics-archive/annual-steel-archive.html, accessed 5 Aug. 2015.

Zini, E. and Scandola, M. (2011) 'Green composites: an overview', *Polymer Composites*, 32(12), pp. 1905–1915.

Index

Italics are used to indicate figures and tables.